MICROSCOPY HANDBOOKS 21

Cryopreparation of Thin Biological Specimens for Electron Microscopy: Methods and Applications

Norbert Roos

EMLB, University of Oslo

and

A. John Morgan

School of Pure and Applied Biology, University of Wales College of Cardiff

Oxford University Press · Royal Microscopical Society · 1990

Oxford University Press, Walton Street, Oxford OX2 6DP

Oxford New York Toronto
Delhi Bombay Calcutta Madras Karachi
Petaling Jaya Singapore Hong Kong Tokyo
Nairobi Dar es Salaam Cape Town
Melbourne Auckland

and associated companies in
Berlin Ibadan

Royal Microscopical Society
37/38 St Clements
Oxford OX4 1AJ

Oxford is a trade mark of Oxford University Press

Published in the United States
by Oxford University Press, New York

British Library Cataloguing in Publication Data
Roos, Norbert
Cryopreparation of thin biological specimens for electron microscopy.
1. Molecular biology. Electron microscopy
I. Title II. Morgan, John A.
574.88028
ISBN 0-19-856424-4

Library of Congress Cataloging in Publication Data
Roos, Norbert.
Cryopreparation of thin biological specimens for electron microscopy:
methods and applications/Norbet Roos and A. John Morgan.
p. cm.
Includes bibliographical references.
1. Cyrobiology. 2. Electron microscope—Technique. I. Morgan,
A. John. II. Title.
QH324.9.C7R66 1990 578'.6—dc20 90-31359
ISBN 0-19-856424-4

Typeset by Cotswold Typesetting Ltd, Cheltenham
Printed in Great Britain by
Biddles Ltd., Guildford & King's Lynn

Preface

This handbook is written primarily for researchers, postgraduate students, and technicians who may have no previous experience of electron microscopical cryotechniques. However, a basic familiarity with biological electron microscopy is assumed. Whenever possible, the reader is directed to available texts in the RMS series for pertinent background information. Otherwise the book is written so that it 'stands on its own feet'. It is not a straightforward 'recipe book': it deals with the essential principles of the subject, but also seeks to provide reliable practical guidelines. We hope that the book will be rather more than an elementary 'starter' and that it will serve as a source of stimulation for transmission electron microscopists who have already experienced both the joys and frustrations of cryotechniques.

Oslo   N.R.
Cardiff   A.J.M.
1989

Acknowledgements

We wish to express our gratitude to Professor Jaques Dubochet, now of the University of Lausanne, who patiently and enthusiastically introduced us to his world of 'beautiful water'. Our grateful thanks also go to those individuals and organizations who contributed illustrative material for the handbook, to Keith Ryan and Tony Robards for their helpful comments on our manuscript, and to all those authors upon the results of whose labours we have copiously drawn. Finally, many thanks to the General Editor of the RMS Handbook Series, Dr Christopher Hammond, for his forbearance during a protracted incubation period, and to Ms Sabina Thompson for typing much of the text.

Contents

Table of Units and Conversions

Prefix	Symbol	Multiplying factor
giga	G	10^9
mega	M	10^6
kilo	k	10^3
centi	c	10^{-2}
milli	m	10^{-3}
micro	μ	10^{-6}
nano	n	10^{-9}
pico	p	10^{-12}

Physical quantity	Name of SI unit	Symbol for SI unit	Definition of unit	Conversion
Electron	e^-			
Electric current	ampere	A	A flow of 1 coulomb per second	
Electric charge	coulomb	C	A definite quantity of electrons, i.e. about $6.25 \times 10^{18}\ e^-$	A.s
Current density	—	$A\,m^{-2}$	Ampere per square metre	$10^4 \times$ (ampere per square centimetre)
Electron dose	—	$C\,m^{-2}$	Coulomb per square metre	$1\,C\,m^{-2} = 6.25\ e^-$ nm^{-2}
Absorbed dose	rad	rad	10^{-2} Joules per kg	$(40\,Mrad \equiv 1\,C\,m^{-2}$ at $100\,kV)$
Pressure	pascal	Pa	$m^{-1}\,kg\,s^{-2}$	7.5×10^{-3} Torr $= 9.87 \times 10^{-6}$ atmospheres $(100\,kPa = 1\,bar)$

1 Water, cells, and electron microscopy

1.1 Introduction

Life on earth could not exist without water; living organisms originated in water, and continue to be biochemically, physiologically, and ecologically dependent on it. Most living cells contain more than 70 per cent water non-uniformly distributed within membrane-limited aqueous domains. The biological importance of water—the only inorganic liquid occurring naturally on earth, and the only compound existing naturally in all three physical states: solid, liquid, vapour—is inextricably linked to its unique and extremely anomalous set of physicochemical properties. Some of the exceptional properties of water are: (i) it has a negative volume of melting, i.e. it expands on freezing, so that ice has a volume about 9 per cent greater than the water from which it formed; (ii) its density is a maximum in the normal liquid range (at 277 K) instead of at the freezing point; (iii) it has anomalously high melting (273 K), boiling (373 K), and critical (647 K) temperatures; (iv) it has the highest specific heat of any known substance except liquid ammonia; (v) it has a high heat of vaporization; (vi) it has a high dielectric constant, which contributes to its role as the universal solvent in biochemical reactions; (vii) its melting point is depressed when it is placed under pressure, an unusual property shared only with bismuth and gallium; (viii) it can exist as numerous crystalline polymorphs—at least nine, including those formed at high pressure. (The high-pressure polymorphs are not of general biological interest.)

Ultimately, all the unusual properties of water must be explained in terms of the intermolecular forces present (hydrogen bonds: ~ 21 kJ mol^{-1}; van der Waals force: ~ 1.3 kJ mol^{-1}). For example, the high boiling point of water results principally from the large amount of energy required to break two hydrogen bonds per molecule evaporated. Methane, with a similar molecular weight to water, boils at 112 K because of the absence of hydrogen bonds and because of the small amount of energy required to break the stabilizing van der Waals forces. Despite the ubiquity and obvious geochemical and biotic significance of water, we are only beginning to obtain a comprehensive molecular understanding of the structure of the liquid (Stillinger 1980). For interested readers, an excellent brief review of the physical properties of liquid water and its solvent effects is provided by Franks (1983).

1

1.2 Electron microscopy of cells

The environment within an electron microscope, with its required vacuum of less than 1.33×10^{-3} Pa (1.0 Pascal = 7.5×10^{-3} Torr = 9.87×10^{-6} atmospheres), is totally inhospitable for fresh biological soft-tissues. The water within the cells and extra-cellular spaces would vaporize explosively if exposed to such low pressures. First it would boil, then the remaining tissue water would freeze due to the removal of the latent heat of vaporization, and finally the frozen water would evaporate. Severe structural damage to the specimen would ensue, caused by the large surface tension forces generated during drying and by the ice crystals formed during the freezing phase.

The modern electron microscope, besides being an optical device with a high-resolution imaging capability, is frequently equipped with one or more detectors permitting the compositional analysis of the observed specimen. Electron microscopy, therefore, offers the biologist the possibility of 'the extension of morphology into biochemistry and the bridging of the gulf between the so-called sciences of matter and the sciences of form' (Needham 1936). Biological electron microscopists have adopted three main general strategies for overcoming the serious constraints arising from the incompatibility of their water-filled specimens with the high-vacuum working conditions within the microscope.

1. The hydrated unfrozen specimen is enclosed within a fluid-filled or vapour-saturated *environmental cell* (Valdrè 1979; Butler and Hale 1981), and viewed by high-voltage electron microscopy. In principle such techniques promise the possibility of observing dynamic cellular activities, but in practice severe optical limitations and radiation damage effects conspire to limit the amount of viable biological data that has emerged from this approach.

2. The tissue specimen is *chemically fixed* so that cellular processes are arrested and cell contents are immobilized by precipitation, denaturation, and cross-linking. Tissue water is subsequently removed by immersion in an organic solvent, and the dehydrated specimen infiltrated with a suitable resin for thin sectioning purposes.

This is the commonest general approach for the morphological study of cells and tissues; detailed procedures and theoretical discussions may be found in the books by Glauert (1974) and Hayat (1981). However, exposure to aqueous organic media inevitably alters both the structure and composition of the tissue samples so that they bear little resemblance to their *in vivo* states. This was recognized as long ago as 1899 by W. B. Hardy, who stated: 'Reagents confer a structure on colloidal matter which differs in most cases in kind, in some cases in degree, from the initial structure. Hence it is inferred that the structure seen in cells after fixation is due to an unknown extent to the action of the fixing reagents' (Hardy 1899). Amongst the more serious artefacts imposed by 'wet chemical' preparative procedures are: the leaching and translocation of soluble molecular species such as proteins, lipids, and sugars (Hayat 1981; Coetzee and Van der Merwe 1984), and ions (Morgan 1980); enzyme inhibition; and the denaturation of antigenic groups (see Chapter 6).

3. The tissue water is immobilized by freezing, i.e. *cryofixation.* This simple statement conceals an enormous technical challenge, and considerable care must be taken before, during, and after cryofixation if the artefacts introduced are to be less damaging and restrictive than those the technique seeks to avoid.

1.3 Cryobiology

Cryobiology, the study of the effects of temperature on biological systems, dates from at least 1663, when Henry Power observed the survival of vinegar eelworms frozen for several hours and then thawed. It has now become a mature discipline with several diverse objectives (Franks 1978; Morris and Clarke 1981). For example, it includes the study of those organisms that have evolved and exploited various biophysical processes for overcoming the stress of low temperatures in their natural environments. These adaptations include cryptobiotic dehydration, the supercooling of body fluids, and the synthesis of anti-freeze molecules (Storey and Storey 1988). Cryobiology is also concerned with the development and application of methods for the cryopreservation of viable blood cells, transplant tissues, sperm, mammalian embryos, seeds, microorganisms, and cultured tissues for many medical, agricultural, and genetic purposes (Morris 1981). In addition, cryobiology includes those studies where cryofixation is used in an attempt to preserve ultrastructural and physiological features within cells and tissues. This latter aspect is, of course, the subject of the present handbook.

Cryotechniques differ considerably depending on the specific aims of an investigation: freezing regimes adopted to maximize cell viability after thawing are fundamentally different from those where cryofixation is the goal (see Fig. 1.1). Whatever the objective, practical methods based on a sound understanding of the underlying principles of the freezing process in aqueous biological systems are preferable to those derived empirically.

1.4 The freezing process: heterogeneous and homogeneous nucleation

Very pure water is unlikely to freeze at temperatures around 273 K. Water which is cooled below its freezing point without ice-formation is said to be in a metastable *supercooled* or *undercooled* state. The degree of supercooling

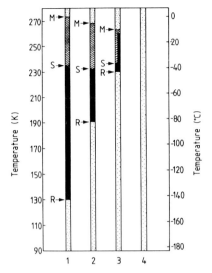

Fig. 1.1. Variation in the freezing and recrystallization characteristics of pure water and different cell types (simplified and redrawn from Robards and Sleytr 1985. 1. *Pure water* melts at 273 K (M), but may be supercooled to about 235 K (S). Recrystallization of ice can occur down to a temperature of ~ 130 K (R). Crystal growth is, therefore, possible during cooling, warming, or storage anywhere in the temperature range M to R. 2. In *active living cells* M and S are lowered and R is raised. The critical temperature range in which ice crystallization can occur is thus reduced. (The addition of solutes to water has a similar effect). 3. In *frost-hardy or cryoprotected cells* M may be further lowered to about 263 K and R raised to about 230 K. Under some favourable circumstances, S may be so low that the cells may solidify without crystallization. 4. In *dry cells* ice cannot form in the absence of water.

depends on the purity of the water: extremely pure water can be supercooled to 235 K at normal atmospheric pressure. Ice crystal formation from the supercooled state is a random event initiated by the nucleation of a single ice crystal. Crystallization leads to the release of latent heat (about 25 kJ mol^{-1}) which warms the water back towards its melting point (Fig. 1.2). When sufficient energy has been removed from the system, the crystals become more stable and spontaneously act as seeds for further crystallization without external assistance. This is termed *homogeneous nucleation*. Further cooling to temperatures around 138 K (the *recrystallization temperature*) will stop ice crystal growth. It is of immense practical significance to note that whatever the original cooling rate, ice crystals will continue to grow in a temperature-dependent fashion at all temperatures down to the recrystallization point. If the energy released by the crystallization process is not removed quickly, the ice crystals formed will be relatively large. Only if the energy generated in the system is removed more quickly than it is released will it be possible to *vitrify* the water, i.e. avoid crystallization (see Chapter 5).

Whereas in very pure water ice crystals themselves act as seeds or

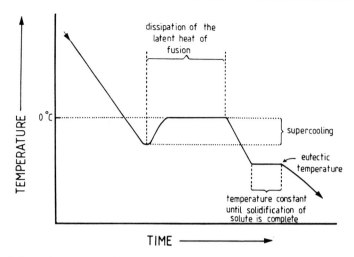

Fig. 1.2. Schematic representation of the freezing process in a dilute salt solution. N.B. Eutectic temperatures for biologically important salts are relatively high e.g. KCl = 262 K, NaCl = 251 K, CaCl = 218 K. Some solutes such as glycerol do not form true eutectics.

catalysts, in other systems insoluble particles such as dust can act as seeds and help to stabilize small ice crystals. This is termed *heterogeneous nucleation*). Structurally, the 'seeds' are thought to mimic the ice crystal surface thus allowing ice growth.

In biological soft tissues the situation is not quite as simple (Fig. 1.3), since the cells consist of a large number of membrane-bound aqueous compartments containing mixtures of inorganic and organic solutes. The 'state' of water within cells and tissues is not precisely known, but it is clear that in a 'typical' cell containing about 80 per cent water the freezing point is depressed by only 2 K below that of pure water, and the recrystallization temperature is *raised* to almost 188 K. The process of ice crystallization in aqueous solutions, including biological tissue specimens, always involves *phase separation*: growing ice crystals draw water from the unfrozen fraction, so that the concentration of solutes in the remaining aqueous phase is increased. This process continues remorselessly whatever the crystal size until the unfrozen salt solution reaches the *eutectic temperature*, i.e. the temperature at which ice is unable to coexist with the concentrated salt solution, at which point the solution freezes (Fig. 1.2). Phase separation during freezing produces high osmotic gradients and pH effects which, together with the physical impact of the growing ice crystals, result in the redistribution of intra- and extra-cellular solutes and the destruction of membranes. In extreme cases such artefacts can easily invalidate structural and analytical studies on cryofixed biological materials.

It has already been mentioned that the methods adopted for freezing cells to maintain viability after thawing, and for the cryofixation of cell ultra-

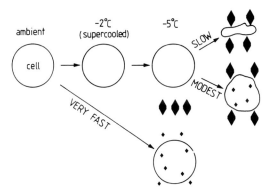

Fig. 1.3. Diagrammatic representation of the effects of different cooling rates on cells. *Slow cooling rates* (< 1.0 K s^{-1}) result in extracellular ice crystallization and allow time for osmotic equilibrium to be maintained by the withdrawal of cell water: the cell shrinks and is deformed, but is frequently viable after thawing. *Intermediate cooling rates* (1.0–1000 K s^{-1}) are too fast to permit mass withdrawal of cell water, so large ice crystals can be formed intracellularly and extracellularly: cell structure and composition is severely damaged and viability is not retained. *Very high cooling rates* (> 1000 K s^{-1}) produces extremely small intracellular ice crystals. (Redrawn from Robards & Sleytr 1985.)

structure and composition, differ considerably. Instructive guidelines emerge from an examination of these differences in general approach (Fig. 1.3). The inescapable conclusion is that *the successful cryofixation of cells for electron microscopic morphological and analytical examination can only be achieved with very high cooling rates*, preferably in excess of 10^4 K s^{-1} (Plattner and Bachmann 1982). Freezing fresh tissue specimens without visible ice crystal damage can, unfortunately, only be achieved with small, thin specimens under favourable conditions (see Chapters 2 and 5). Minimizing ice damage in larger specimens yields preparations whose superficial layers (only) approximate the *in vivo* state, and which are more readily sectioned than severely ice-damaged preparations. The incentives for optimizing freezing conditions are self-evident.

1.5 Cryotechniques for the preparation of thin biological specimens for electron microscopy: scope and advantages

Cryofixation is the first step and is the major limiting factor determining whether or not a tissue specimen can be brought to the microscope in a high-fidelity state. This physical procedure clearly offers several outstanding advantages over chemical fixation:

1. Cells may be processed and brought to the microscope surrounded by their native extracellular fluid environments.

2. Enzymes and antigens are not denatured.

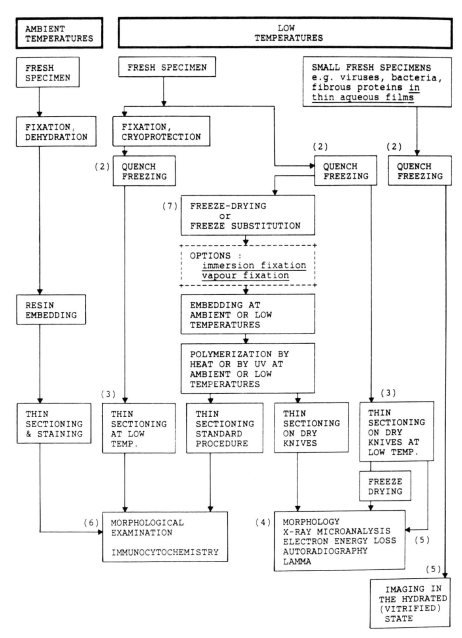

Fig. 1.4. Diagram showing the range of different methods of processing fresh biological specimens for examination in the transmission electron microscope. (Numbers refer to the Chapter numbers where a method is discussed.)

3. Cells are rapidly immobilized (within 0.1–1 ms) (Plattner and Bachmann 1982). 'Time-resolved' electron microscopic studies are therefore possible, i.e. dynamic cellular activities involving fast molecular and membrane rearrangements, such as muscular contraction and exocytosis, can be intercepted and preserved for examination.

4. Freezing physically hardens the specimen so that it may be sectioned without infiltration with extraneous materials.

Cryofixed biological tissues must be further manipulated before they can be examined in the electron microscope. Specimen integrity can all too easily be seriously compromised during these various manipulative procedures, *including during electron irradiation in the microscope.* It is, therefore, absolutely essential to consider each individual cryopreparative procedure, whatever the particular biological objectives, as a firmly linked chain of interdependent activities. Specimen damage accumulates, and certainly cannot be reversed, during passage along the cryo-chain!

We have tried to meet three objectives in this handbook:

1. To provide firm practical guidelines to those cryotechniques where the cryofixed specimen is sectioned or otherwise presented as a thin film for examination in the transmission electron microscope. Particular attention is devoted to those techniques proving to be of considerable contemporary (and, we believe, future) interest (Fig. 1.4). Some of the techniques described may also be eminently suitable for the high-fidelity preparation of biological specimens for examination by other optical and microprobe techniques (see Hall and Gupta 1983).

2. To persuade the reader that cryoelectron microscopy has 'come of age' and has much to offer the cell biologist. There exists a broad understanding of some of the crucial biophysical events involved in the freezing of biological specimens, coupled with a wealth of accumulated practical expertise in several independent laboratories. Many of the pitfalls in cryopreparation that we mention can be circumscribed (but do not ignore them—they are unforgiving!), and the techniques described can be learnt and routinely used in any laboratory that possesses the appropriate complement of commercially-available equipment.

3. To stimulate and guide the reader to design and make inexpensive, but effective equipment that can accomplish a lot in areas such as plunge and slam freezing, freeze-substitution, and freeze-drying.

2 Freezing methods

2.1 Introduction

The rate at which a specimen is cooled is a crucial parameter, since it determines the 'quality' of the frozen specimen. In most situations it is necessary to cool very rapidly to arrive at or below the recrystallization point of the biological system in the shortest possible time in order to:

(1) avoid or at least minimize the formation of ice crystals which form in the absence of cryoprotectants (Stephenson 1956 and 1960);

(2) maintain the original distribution of diffusible components (Roos and Barnard 1986; Zglinicki *et al.* 1986);

(3) arrest transient morphological changes which are sometimes faster than the cooling rates required to prevent ice crystal growth (Heuser *et al.* 1979; Plattner and Knoll 1987).

Successful physical fixation (freezing) permits high 'spatial' resolution for morphological studies and high 'time' resolution for analytical studies (X-ray microanalysis) and studies of dynamic processes. This cannot be achieved by chemical fixation (Zingsheim and Plattner 1976; Plattner and Bachmann 1982). However, if cooling rates are not high enough, biological samples often separate into domains of biological material, concentrated solutes with some residual water, and pure crystalline water (ice crystals). In the worst cases the crystals can destroy structures as large as organelles. But even if ice crystal damage is not obvious, displacement of, for example, solutes could have taken place, because the formation and growth of ice crystals requires the movement of water.

Detailed guidance on the theory and practice of freezing can be found in the books by Bald (1987) and Robards and Sleytr (1986). Important questions that need to be addressed prior to freezing a given specimen are: (i) what is the nature/size of the specimen? (ii) can the specimen be manipulated prior to freezing without compromising its structure and composition? (iii) what are the specific objectives of the study?

2.2 Sample handling prior to freezing

2.2.1 Sample handling in general

Keeping the sample as close to the *in vivo* state as possible is paramount. For example, the quantitative analysis of diffusible elements in cryosections by

9

means of electron probe X-ray microanalysis (EPXMA) requires extreme care to avoid accidental redistribution at every stage of specimen preparation and subsequent analysis.

It is easily forgotten that specimen preparation starts before quench-freezing. The handling of an animal and the use of anaesthetics can lead to stress-related changes in its neural and hormonal status. Dissection of samples can lead to mechanical deformation, reduced oxygen supply, and falling blood pressure. As a result, gross post-mortem changes can occur resulting in disturbed element distributions in the cells (Somlyo *et al.* 1977; Zglinicki *et al.* 1986). It is, however, possible to minimize the risk of pre-freezing artefacts. In cases where the animal has to be dissected, the animal and/or the exposed tissue should be located in a humidity chamber with temperature control. If the tissue or organ is to be taken out, it should be placed in a bath containing aerated physiological medium (Hagler and Buja 1984). To avoid post-mortem changes totally, dissection and freezing should be simultaneous. To date this is only possible by:

(1) quench-freezing the tissue *in situ* against two cold copper blocks attached to pliers (Hagler and Buja 1984) operated either manually or automatically;

(2) a clamping device holding two opposing melting Freon popsicles (Somlyo *et al.* 1985);

(3) cryoballistic methods that use a hypodermic needle which is shot through the tissue. The cylindrical sample is frozen by the cold needle and falls into a cryogen bath (Monroe *et al.* 1968; Hearse *et al.* 1981; Zglincki *et al.* 1986).

2.2.2 Use of cryoprotectants

When bulk specimens are frozen, the thermal conductivity of water and/or ice will limit the rate of cooling within the specimen. As a result, only the specimen surface adjacent to the coolant will be well frozen (with the possible exception of tissues frozen under high pressure), thus limiting morphological and X-ray microanalytical studies to a small fraction of the sample volume. Furthermore, in excised specimens the well-frozen surface region will also comprise cells which have been severely damaged during tissue sampling. The presence of ice crystals will make cryosectioning more difficult, reduce spatial resolution, and lead to redistribution of electrolytes. To avoid ice crystal damage during freezing of bulk specimens, the use of cryoprotectants (chemical antifreeze agents) has been proposed (Skaer 1982). Cryoprotectants are agents that:

(1) lower the temperature at which homogeneous nucleation occurs;

(2) raise the recrystallization temperature;

(3) reduce the 'free' water in the system.

Cryoprotectants can conveniently be divided into two groups: (i) those that penetrate cells and subcellular compartments, e.g. glycerol, dimethyl-sulphoxide, and ethylene glycol, and (ii) those that do not, e.g. poly-vinylpyrrolidone, dextran, and sucrose. They should neither damage the cells nor be toxic to them. Penetrating cryoprotectants cannot be used where the original distribution of electrolytes or other soluble components must be preserved, since they alter the properties of cell membranes, with a con-sequent redistribution or loss of solutes. Even though non-penetrating cryoprotectants interfere less with cellular functions such as locomotion, volume maintenance, and excitability, the concentrations required to obtain cryoprotection (>20 per cent) are very high and do have varying effects on different tissues (Barnard *et al.* 1984). The use of a cryoprotectant must therefore be considered carefully, and one must be aware of the possible physiological and structural artefacts.

2.3 Freezing

2.3.1 Cooling rates

The rate of cooling, the most important parameter in biological freezing, is a function of the composition, size, and shape of the specimen, the nature of the specimen holder, the rate of entry of the specimen into the cryogen, the depth of plunge, and the properties of the cryogen. Measured cooling rates will, in addition, be dependent on the nature and size of the measuring device (thermocouple). It is therefore almost impossible to compare cooling rate measurements reported from different laboratories unless additional infor-mation such as specimen thickness, ratio of specimen volume to thermo-couple volume, and rate of entry into the cryogen are given (Ryan *et al.* 1987) (Table 2.1).

2.3.2 Coolants

If a specimen is to be frozen by immersion into a liquid cryogen one has the choice between a number of different liquids. To make sure that gas formation at the specimen–coolant interface (film boiling) is minimal and heat transfer maximal, liquid cryogens are selected for low melting points and high boiling points. The formation of a stable gas layer around the specimen tends to insulate it from the coolant; for this reason liquid N_2 is not an efficient primary cryogen for critical high-resolution electron micro-scopical studies of non-cryoprotected tissues. In addition a suitable coolant should:

(1) have good heat conduction and high heat capacity;

Table 2.1. *Cooling rates for four different cryogens* (K s^{-1})

Propane	Ethane	Freon 22	Liquid nitrogen	Thermocouple diameter (μm)	Entry velocity (m s^{-1})	Reference
8892	12109	5950	569	300	1.12	Ryan et al. 1987
9787	13038	6706	731	300	2.25	Ryan et al. 1987
1647	2103	1138	—	550*	1.12	Ryan et al. 1987
1858	2201	1276	—	550*	2.25	Ryan et al. 1987
19100	—	9000	800	300	1.4	Elder et al. 1982
98000	—	66000	16000	70	0.49	Costello and Corless 1978

All rates measured between 273 K and 173 K.
*Thermocouple (300 μm) surrounded by a 125 μm layer of hydrated gelatine.

(2) have high fluidity at the low temperatures required;

(3) have a high density;

(4) be safe to use;

(5) be inexpensive.

A list of cryogens is given in Table 2.2. Propane and ethane fulfil the requirements extremely well (Ryan *et al.* 1987) but are *highly explosive* and need to be handled with considerable care (see Chapter 8).

Table 2.2. *Characteristics of the most commonly used liquid cryogens*

Liquid	Melting point (K)	Boiling point (K)	Specific heat near melting point (J/g/K)	Thermal conductivity near melting point (mJ/m/s/K)
Ethanol	156	352	1.9	206
n–Pentane	143	109	1.9	177
Halocarbon 12	115	243	0.8	138
Halocarbon 22	113	232	1.1	152
Isopentane	113	301	1.7	182
Ethane	90	184	2.3	240
Propane	84	231	1.9	219
Liquid nitrogen	63	77	2.0	153
Liquid helium	1.7	4.2	4.5	18

Data extracted from Robards and Sleytr 1985

2.3.3 Freezing techniques

There are seven main rapid freezing techniques available at present. These are summarized in Table 2.3. The user would obviously prefer a freezing method that is fast (to avoid post-mortem changes), simple, reproducible, inexpensive, and economic during operation. Since some of the methods are quite complex and require expensive devices—for example, helium-cooled copper mirror freezing (Escaig 1982) and high pressure freezing (Moor, 1987)—or are not widely used—for example, excision freezing (Hearse *et al.* 1981)—we would like to restrict ourselves to the description of a few 'simple' freezing techniques.

2.3.3.1 Immersion freezing

The specimen is plunged manually or mechanically into a liquid or melting cryogen. The specimen can be a bulk sample, which should be as small as possible, mounted, for example, on a metal holder (Zasadzinski 1988). Alternatively, it could be an aqueous suspension (containing, for example, a high density of viruses) mounted as a thin film on a coated or uncoated electron microscopical grid (Chapter 5). In both cases care must be taken that the specimens do not dry out prior to freezing, that they are not allowed to pre-cool in the cold gas phase above the cryogen bath, and that the rate of

Table 2.3. *Comparison of available rapid freezing methods*

Method	Description	Freezing	Costs	References
Plunge-freezing	Specimen is plunged manually or mechanically into liquid or melting cryogen (e.g. subcooled nitrogen, propane, ethane, fluorocarbons etc.)	Adequate for most requirements	Inexpensive, easily home made	Barnard 1982 Elder *et al.* 1982 Ryan *et al.* 1987
Spray-freezing	Spraying of droplets of suspensions into a suitable cyrogen	Extremely good since specimen volume small	Inexpensive, workshop	Bachmann and Schmitt 1971 Dubochet *et al.* 1982*a*
Jet-freezing	Spraying of a jet of cold liquid cryogen on to one or both sides of a stationary specimen, which is sandwiched between thin metal plates	Good freezing of cell culture monolayers	Relatively expensive	Moor *et al.* 1976 Müller *et al.* 1980 Pscheid *et al.* 1981 Knoll *et al.* 1982
Cold-block freezing	a) Metal blocks attached to the jaws of pliers are used to freeze tissue *in situ* without prior excision	Good freezing in two thin surface zones	Very inexpensive	Ingram and Ingram 1984 Hagler *et al.* 1984 Roos *et al.* 1989
	b) Specimen is impacted on to a liquid nitrogen or liquid helium cooled metal mirror	Extremely good freezing deep into the tissue	Expensive when run with helium	Heuser *et al.* 1979 Escaig 1982 Philips and Boyne 1984
High-pressure freezing	Freezing at high (2000 bar) pressures to subcool the water. Cooling rates can therefore be reduced	Good freezing of even large samples (1 mm^3)	Expensive	Moor 1987
Excision freezing	A cold needle is plunged into the tissue thus cooling and dissecting the tissue simultaneously	Not very convincing	Not very expensive	Hearse *et al.* 1981 Von Zglinicki *et al.* 1986

entry is high enough. The liquid cryogen should be cooled close to the solidification point and stirred continuously in order to avoid temperature gradients. The specimen has to pass through a large volume of coolant to ensure maximal heat transfer. A simple mechanical device for achieving these objectives in thin-film preparations mounted on grids, and which can be 'home-made' with modest workshop facilities, is presented in Fig. 2.1. The cryofixation of suspensions with ethane or propane as cryogens can lead to vitrification of the sample (Chapter 5).

The cryofixation of non-cryoprotected bulk specimens with this device would be relatively poor. However, more elaborate devices can provide: sufficiently high, variable, but reproducible entry velocities in the range $1-10$ m s^{-1}; a depth of coolant sufficient to permit the specimen to keep moving at its entry velocity for the duration of the early phase of the critical cooling range (effectively $273-173$ K); coolant stirring; and a means of

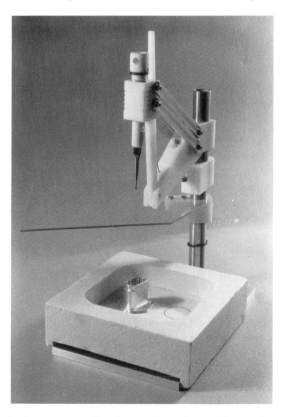

Fig. 2.1. Simple device for plunge-freezing small specimens. The specimen is held between very fine tweezers to minimise heat input. Upon release gravity will pull the arrangement into the coolant. (Whilst this device is suitable for freezing thin films on grids, it would not produce a high enough entry velocity and depth of plunge into the coolant to achieve high cooling rates in larger specimens.)

preventing pre-entry freezing in the vapour phase. These are relatively easy to build in a workshop—see a range of designs described by Robards and Sleytr (1986)—or can be purchased from commercial sources, for example, Balzers and Reichert-Jung (Cambridge Instruments).

2.3.3.2 Spray freezing

The spray freezing technique has been developed to reduce the size of the sample and thus the time needed to freeze it (Bachmann and Schmitt 1971; Plattner *et al.* 1973). Usually an atomizer is used to produce and shoot small droplets of a suspension into liquid propane. The propane is subsequently evaporated at 188 K using a rotary pump. The droplets are very small and the technique can therefore only be used to freeze suspensions of cell fractions, isolated cells, unicellular organisms, and viruses. The freezing quality is extremely good.

2.3.3.3 Jet freezing

The principle of the two freezing techniques described above is to move the specimen through a coolant at high speed to ensure that the coolant is not warmed up locally and that the specimen remains in contact with the 'renewed' coolant at a given low temperature ($\leqslant 90$ K). The alternative approach is to keep the specimen in a fixed position and have the cryogen flow rapidly past it. A jet of cryogen (usually nitrogen-cooled propane) hits the specimen, which for freeze-fracture is sandwiched between two grids, either from both sides as a double jet (Moor and Müller 1976; Müller *et al.* 1980) or from one side as a single jet (Pscheid *et al.* 1981; Knoll *et al.* 1982). When using the single jet technique the side of the specimen away from the jet has to be insulated to minimise heat input from this side (Fig. 2.2).

2.3.3.4 Freezing on cold metal mirrors

The high thermal conductivity and diffusivity of some metals (e.g. copper and silver) inspired attempts as long as three decades ago (Eränko 1954) to freeze biological material by impacting against the surface of cold, highly polished metal blocks. A series of freezing devices emerged using metal blocks cooled by liquid nitrogen (Van Harreveld *et al.* 1974) and liquid helium (Heuser *et al.* 1979; Escaig 1982). The resulting specimens possess a very well-frozen flat surface. However, the depth of the well-frozen layer is limited by the low thermal conductivities of ice and the specimen itself.

A very simple cold block freezing device (Hagler and Buja 1984; Ingram and Ingram 1984) is shown in Fig. 2.3. Two highly polished copper blocks are attached to the jaws of a pair of pliers by silver solder. The two insert surfaces are adjusted to be perfectly parallel with a gap of 0.2–0.3 mm when the pliers are closed. The pliers can be cooled in liquid nitrogen, brought to

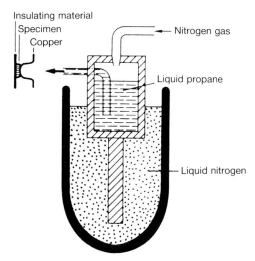

Fig. 2.2. Apparatus for propane jet-freezing a specimen from one side only. The rapid withdrawal of heat from one side, coupled with a thermally insulating backing on the other side to reduce heat input during freezing, ensures that high cooling rates can be achieved with this simple arrangement. (Redrawn from Pscheid *et al.* 1981.)

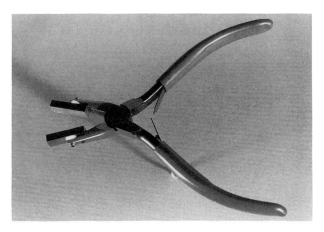

Fig. 2.3. An extremely simple cold-block freezing device. Two highly polished copper blocks are attached to the jaws of a pair of pliers. Two adjustment screws ensure that the two surfaces are parallel. Pliers are cooled in liquid nitrogen in the closed position in order to avoid condensation of water onto the surfaces. The device allows tissue specimens to be frozen *in situ*.

the specimen in the closed position, then opened quickly to grab and squeeze-freeze the specimen. The result is a thin wafer of tissue with two well-frozen, flat, and parallel surfaces. The wafers are very convenient for subsequent cryosectioning. This freezing device allows freezing of tissue *in situ*. Since the pliers are operated by hand, the reproducibility is somewhat low compared to the other cold metal mirror freezing techniques. A mechanically operated pair of pliers (the 'Model 669 Cryosnapper') has recently been introduced (Gatan), but not much information about freezing quality and reproducibility is available to date.

2.3.4 Evaluation of freezing quality

Vitrification of the whole specimen is possible under some circumstances (Dubochet *et al.* 1982*c*), but the specimen needs to be very small or thin. For most specimens vitrification is unfortunately limited to a thin surface layer. The goal must be to reduce the ice crystals to a size that does not affect the intended observations. In the case of X-ray microanalytical measurements, the magnitude of the ice crystals and hence the displacement of elements should be smaller than the required analytical spatial resolution (e.g. <0.1 μm) in the area of interest. There are different ways of checking the quality of the freezing procedure (cryofixation). The tests are mostly indirect and designed to help evaluate the usefulness of freezing techniques.

1. A standard solution can be frozen and examined for ice crystals by freeze-fracture methods, using, for example, 5% glycerol (Plattner and Bachmann 1982) or a dilauryllecithin–water system (Costello *et al.* 1982).

2. The water in frozen samples can be substituted with a solvent at low temperatures (freeze-substitution); the size of the ice crystal 'ghosts' is then measured after plastic embedding and sectioning. Even though it has been shown that in some biological systems the recrystallization process is relatively slow (Steinbrecht 1985), freeze-drying or freeze-substitution methods are possibly subject to rearrangements of materials and liable to secondary ice crystal formation. They are therefore of limited use for testing of the state of water in the sample.

3. A direct, reliable, and unambiguous test is to record an electron diffraction pattern (Dubochet and McDowall 1981). Such a diffraction pattern allows identification of vitrified ice, cubic ice, and hexagonal ice (Fig. 2.4) in thin specimens such as fully hydrated cryosections (Chang *et al.* 1983; Griffiths *et al.* 1984) and quench-frozen films of suspensions (Adrian *et al.* 1984; Dubochet *et al.* 1985) (see Chapter 5).

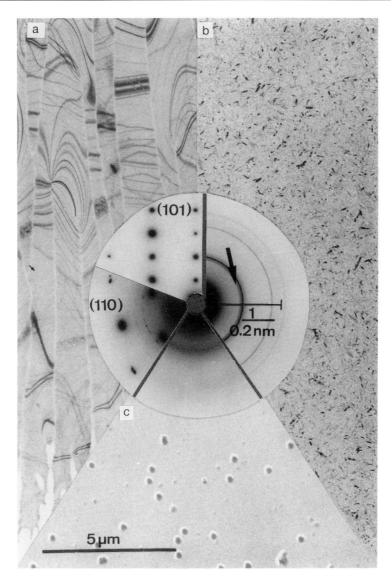

Fig. 2.4. Typical images and electron diffractograms of the three forms of ice. The direct images are all presented at the same magnification ($\times 7500$). (a) Hexagonal ice, I_h, obtained by the rapid freezing of a thin water layer spread on a carbon film; the thickness of the layer shown in the micrograph is 50–80 nm. The corresponding diffractograms, which were taken from other specimens, show the (110) and (101) planes. (b) Cubic ice, I_c, obtained by warming a layer of I_v. The small contribution of the (100) form of I_n has been marked on the diffractogram (arrow); the I_c layer is approximately 70 nm thick. (c) Vitrified ice, I_v, obtained by deposition of water vapour in the electron microscope on a film supporting polystyrene spheres as a focusing aid; the ice layer is approximately 70 nm thick. The shadowing effect seen in this preparation demonstrates that the flux of water molecules hitting the specimen in the microscope was anisotropic. (With permission from Dubochet *et al.* 1982*a*.)

3 Cryoultramicrotomy

3.1 Conventional ultramicrotomy

Soon after the introduction of the electron microscope, first attempts to obtain thin sections were made by Von Ardenne (1939). This was an essential requirement due to the small depths of penetration of the electron beam. Sections needed to be ≤100 nm thick, which was one-tenth of what could be achieved with microtomes or cutting devices for light microscopical examinations. Research in the field of thin and ultra-thin sectioning led to the development of microtomes with rotating razor blades to section paraffin embedded objects (O'Brien and McKinley 1943), new embedding techniques (Newman *et al.* 1949), sectioning on to liquid surfaces (Hillier and Gettner 1950) and the introduction of glass knives (Latta and Hartmann 1950). As a result the 'modern' ultramicrotomes emerged, all of which have certain features in common.

3.2 Ultrathin sectioning of specimens at low temperatures

Interestingly enough, the concept of cutting frozen thin sections came along with the early development of sectioning techniques for electron microscopy. In the late 1940s and early 1950s first attempts were made to cut frozen tissue (Fernández-Morán 1952). This early interest in 'cryosectioning' arose because it offers two major advantages over conventional sectioning:

1. Conventional procedures including chemical fixation, dehydration, etc., take several days before sectioning and examination in the microscope can take place. Cryofixation, cryosectioning, and examination in the microscope is achieved in one day.

2. Chemical and structural integrity of the tissue is more faithfully preserved and cryosectioned biological tissue should therefore be ideal for histochemical and immunocytochemical reactions, for autoradiography of soluble compounds, for X-ray microanalysis of electrolytes, and for imaging of fully-hydrated specimens.

In the early days of cryomicrotomy, however, it proved very difficult to obtain usable thin cryosections even after a light chemical fixation. Most electron microscopists therefore preferred to work with plastic-embedded samples.

3.2.1 The modern ultramicrotomes

We will describe the features of a 'typical' modern microtome. Most of these features are basically the same for all the commercially available ultramicrotomes and cryoultramicrotomes.

An ultramicrotome has a knife-stage which can hold a knife of steel, glass, or diamond at an adjustable angle. The knife can be rotated to allow for trimming of the specimen. The specimen is attached to an arm which moves downwards past the knife edge. Between two cutting cycles the specimen arm can be advanced in increments, either mechanically or thermally ('mechanical feed' or 'thermal-feed'). The advance of the specimen should give a linear, reproducible movement. This leads to the detachment of a section of given thickness. The section floats off onto the surface of a liquid (usually distilled water) contained in a trough attached to the knife. Dry sectioning can be performed by simply abandoning the trough liquid. This, however, usually produces folded and wrinkled sections of uneven thickness. If the specimen were allowed to pass the knife edge during the return stroke, the block face would in most cases pick up the sections again. To overcome this problem the specimen arm is deflected backwards or the knife is retracted during the return stroke.

A very important sectioning parameter is the cutting speed, i.e. the speed at which the specimen passes the knife. It is adjustable but held constant during one pass. Cutting speeds are usually quite slow and it would be very time-consuming to perform the return stroke and the approach of the specimen to the knife at the same slow speed. The cutting cycle is therefore divided into two phases: a slow cutting phase, and a rapid return phase. In order to achieve reproducibly uniform serial sections, and reduce the chances of the operator creating mechanical disturbances such as vibrations, the movements of the specimen are automated (with the option of manual operation).

To facilitate the mounting of the specimen, the observation of the sectioning process, and the section collection, all ultramicrotomes provide for adequate lighting and a binocular microscope.

3.2.2 Basic features of a cryoultramicrotome

A good reliable cryoultramicrotome has all the functions described in the paragraph above. In addition it should be possible to control the temperature of (i) the specimen, (ii) the knife, and (iii) the atmosphere around the specimen-knife assembly down to temperatures as low as 110 K without affecting the general performance of the ultramicrotome (i.e. reproducible advance, cutting speed, etc.). For convenience these temperatures should be controlled independently, and set temperatures should be stable over several hours with only small deviations ($\leq \pm 2$ K).

The specimen is usually rather small due to the demands of the freezing process (see Chapter 1). It can either be mounted in the fresh state on a metal stub for plunge-freezing, or it may be frozen in, for example, cryo-pliers and mounted under liquid nitrogen in the jaws of a clamping specimen holder. A cryoultramicrotome chamber should provide enough working space around the knife for (i) the easy insertion of the specimen support, (ii) the easy collection of sections, and (iii) the temporary storage of grids prior to and after section retrieval. Where the transfer of the frozen sections to other working stations (freeze-dryer, coating unit, microscope) is desirable, the chamber should also provide enough space to accommodate these transfer devices. Throughout the entire pathway—from mounting the specimen, to sectioning, to section collection, and to section transfer—the sample should be protected from contamination and from uncontrolled warming, which could lead to thawing or inadvertent freeze-drying. The requirements described above are met by most of the modern cryoultramicrotomes on the market.

3.2.3 The evolution of cryoultramicrotomes

The early approach to cryoultramicrotomy was to put a conventional ultramicrotome into a deep freeze (Bernhard 1965). Improved versions (Slee Cryoultramicrotome) of this so-called 'cryostat' (Appleton 1974) are still in use. The limiting factors for the use of these bulky machines are:

(1) the relatively limited temperature range (the lowest temperature achievable is 170 K);

(2) the temperature is the same for knife and specimen, and adjustments can take a long time because the temperature of the whole assembly has to change;

(3) the knife area is not easily accessible; the operator has to stand whilst manipulating the microtome;

(4) it cannot be used as a conventional ultramicrotome.

The advantages of these machines are the space available for auxiliary equipment and the long running time. Once the system is cooled down it can operate for days. However, for the reasons mentioned above the second approach to cryoultramicrotomy is now more widely used than the cryostat method. By 1966 people had already begun to experiment with cryochamber attachments for existing conventional ultramicrotomes. First reports were published shortly afterwards, and two basically different cryoattachments evolved. Both followed the principle of putting an insulated 'box' or cryo-enclosure around the knife stage and specimen arm of a conventional ultramicrotome in order to keep the temperature in this area stable at a selected, preset value. Whereas Dollhopf and Sitte (1969) designed the specimen arm for maximum mechanical stability and, therefore, extended it through the wall of the cryo-enclosure, Christensen (1971) chose to traverse the wall of the cryo-enclosure with an insulated L-shaped and later U-shaped bridge (Fig. 3.1). Cooling was achieved in both instrument types with a continuous flow of dry cold nitrogen gas which prevented heat and humid air from entering the chamber. The temperatures of the specimen arm and the knife could be regulated independently of each other. Depending on the nature of the specimen, a small temperature differential between the specimen and knife can be an advantage, although this is not generally agreed. These cryochambers are more economic than 'cryostats' since only a small fraction of the ultramicrotome bulk has to be cooled. Furthermore, the ultramicrotome can be used in the conventional way at ambient temperatures because the cryochamber attachment can be removed easily. All the moving parts of the ultramicrotome operate at room temperature, the fine controls for cutting speed, feed, etc., are therefore more reliable and reproducible.

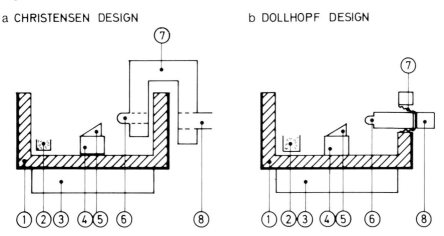

Fig. 3.1. Schematic drawing of the Christensen type (a) and Dollhopf type (b) cryoattachment; 1 = insulating 'box', 2 = liquid nitrogen tank, 3 = knife stage of the microtome, 4 = knife holder, 5 = knife, 6 = specimen, 7 = specimen holder, 8 = specimen arm of the microtome.

The first generation of commercially available cryoattachments (e.g. the LKB 14800 Cryokit) were relatively difficult to use since temperatures were not stable and the cryoattachments were poorly insulated. Contemporary cryomicrotomes have overcome these difficulties.

3.2.4 'Modern' commercially available cryoattachments

For the reasons mentioned above, the cryoattachment is predominant today and is more attractive for the newcomer to the field of cryoultramicrotomy. We will therefore limit ourselves to the description of the commercially available cryoattachments of the three major ultramicrotome manufacturers. The three producers are LKB Produkter AB (Cambridge Instruments), Reichert–Jung (Cambridge Instruments) and RMC (formerly Sorvall). The cryoattachments of the different manufacturers will only fit their own ultra-microtomes.

A universal cryosectioning attachment compatible with all three major ultramicrotomes has become available recently and is produced by RMC (Tucson, Arizona, USA).

3.2.4.1 The LKB system

The CryoNova cryochamber is made from high-density polyurethane and can be attached to the body of the ultramicrotome called 'Ultratome Nova'. A modified version that fits the Super Nova ultramicrotome has been intro-duced recently (Fig. 3.2). The Christensen-type bridge is used to attach the specimen to the specimen arm. The bridge material is low thermal con-ductivity ceramic, with an aluminium head where the specimen holder fits in. The bridge parts outside the cryochamber are heated to offset loss of energy into the cryochamber. The temperatures of the knife and the specimen holder can be regulated independently and can reach a lower limit of 123 K. Three sensors check the knife, specimen, and chamber temperatures and the

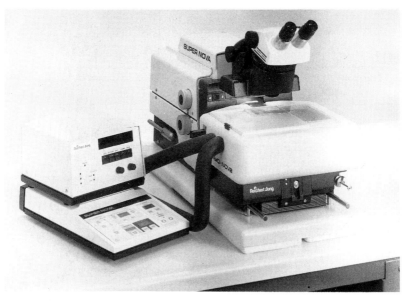

Fig. 3.2. Photograph of the Reichert–Jung (Cambridge Instruments) Super Nova microtome with the Cryo Nova cryoattachment.

values can be recorded using a chart recorder. The cryochamber is automatically filled with nitrogen gas which passes through a gas phase separator to avoid turbulence in the working area. A foil heater ensures a controlled flow of cold nitrogen gas within the cryochamber. The cryochamber is covered by a transparent lid. Sensors in the liquid nitrogen storage tank signal the level to a control unit. A whole set of tools for section and grid handling and grid retrieval is available. However, the tools available are not compatible with existing cryo-transfer units. Cutting at extremely low temperatures—for example to obtain fully hydrated sections—causes ice to build up around the bridge, thus affecting sectioning reproducibility.

3.2.4.2 The Reichert–Jung system

The Reichert–Jung cryosystem FC 4E has been developed and designed for use with the Ultracut E microtome (Fig. 3.3). The cryochamber consists of a three part aluminium tank with styrofoam insulation surrounded by another aluminium wall. This outer wall is kept at room temperature to prevant condensation and ice build-up. The chamber is attached to the knife stage. The two outer parts of the aluminium tank (the 'twin tanks') are connected by a pipe and automatically filled with liquid nitrogen to a given level. This process is controlled by diodes. Nitrogen continuously evaporates from the tank and is guided by the inner chamber wall into the sectioning chamber, creating a laminar upward flow of cold nitrogen gas, which cools the specimen and the knife. The system ensures a stable temperature gradient

Fig. 3.3. Photograph of the Reichert–Jung Ultracut E microtome and the FC 4E cryo-attachment.

from the bottom of the chamber to the open top. The temperature of the nitrogen gas arriving at the bottom of the chamber is 83 K. On its way to the top of the chamber it will warm up to about 123 K. These temperatures are too low for most applications. Separate heaters in the knife holder and in the bridge (Christensen-type alloy steel bridge) close to the specimen, and a special heating plate to heat the chamber gas, allow the temperatures of the knife and the specimen to be controlled independently. The heating of the elements is counteracted by the cold gas and after an initial equilibration period the system runs under steady-state conditions. The system has a front entry for a cold transfer device from Zeiss. All operations are performed without a lid and the cryoattachment is mechanically stable.

The heating of the outer walls is reflected in a high liquid nitrogen consumption (in the latest models nitrogen consumption is somewhat reduced) and the system has to be 'baked out' after use to avoid condensation of water on vital parts of the cryochamber.

3.2.4.3 The RMC system

The cryoattachment produced by DuPont–Sorvall was called the FS 1000 and was designed to fit the Sorvall MT2, MT-2B, MT-2C, and MT5000 ultramicrotomes. RMC produces a cryoattachment called CR2000 which fits all the Sorvall ultramicrotomes and the RMC ultramicrotomes MT6000 and MT6000XL (Fig. 3.4). It is a thermally insulated aluminium chamber attached to the base of the knife holder of these ultramicrotomes. In common with the other cryoattachments it uses the Christensen-type bridge to

Fig. 3.4. Photograph of the RMC MT6000XL microtome with the CR2000 cryoattachment.

support the specimen. The temperatures of the knife and the specimen can be preset digitally within a tenth of a degree, and a temperature control unit registers and maintains the given temperatures very accurately, to a lowest possible temperature of 110 K. Liquid nitrogen is fed into a small tank in the chamber via a phase separator, which prevents nitrogen gas from entering the chamber, and thus reduces turbulence in the chamber atmosphere. A low level function for nitrogen filling can be preset to minimize nitrogen consumption during sectioning at temperatures above 180 K. A heater element in the bottom of the nitrogen tank generates a constant flow of cold nitrogen gas through the chamber to prevent moist air entering the chamber, which in addition is covered with a transparent lid. The cryochamber contains a grid handling device to bring the grid close to the knife edge for section collecton and a storage device for grids. The latest version of the CR2000 has a rotating knife stage ($\pm 8°$) to facilitate trimming.

The nitrogen feed line enters the cryochamber from the top (left side) through the lid, and its position is rather inconvenient.

All three cryoattachments discussed above are provided with standard specimen holders which allow for a variety of specimen support pins to be fitted. For work with flat samples vice-type chuck holders are also provided. The knife holders accept steel, glass, diamond, or sapphire knives.

3.2.5 Specimen preparation for cryosectioning

In the case of tissue 'cubes', sectioning can be facilitated by trimming into a pyramid shaped block. If freezing is performed by slamming the sample against a cold metal mirror the result is a flat, thin, frozen piece of tissue which can either be glued to a metal pin using organic solvents with low melting points (e.g. butylbenzene, 185 K; hexane, 175 K) or clamped between the jaws of a vice-type holder that fits into the specimen arm. Freezing on small metal supports, such as pins, has the advantage that very small volumes of, for example, bacterial suspensions can be frozen. Very often, however, one faces the problem that the sample cracks upon warming up from liquid nitrogen temperature to the sectioning temperature ($T = 50$ K) because of the different thermal expansion coefficients of the biological material and the metal support. The cracks can render sectioning difficult if not impossible. It is also important to recognize that using low temperature 'glues' could compromise subsequent analytical measurements of electrolytes, since the glues have to be applied as liquids, and can therefore lead to local thawing or at least partial warming of the sample. Frozen samples can be stored indefinitely under liquid nitrogen prior to sectioning; they are then transferred to the cryoattachment in a small container filled with liquid nitrogen. Warming of the sample during transfer should be avoided since it might lead to thawing, subsequent redistribution of diffusible elements, and local freeze-drying.

3.2.6 The knife

Once the specimen preparation prior to sectioning is optimized, the most critical parameter for obtaining good thin cryosections is the quality (sharpness) of the knife (Griffiths *et al.* 1983). In most cases this will be a glass knife, since manipulation of the specimen block and section collection from the dry knife might easily result in damage to a diamond knife. Cryosections tend to stick more tenaciously to diamond knives than to glass knives. The ideal knife is broken from a perfectly square piece of glass and the break has to be exactly diagonal to obtain the sharpest knives. This is only guaranteed if the corner edge of the glass square is perpendicular to the plane of the square. To get a glass square that fulfils these requirements one first breaks rectangular pieces of glass exactly twice the length of the square. The resulting symmetrical weight distribution and the symmetrical arrangement of the score line, the breaking pins, and the support studs usually gives good squares (for details see Griffiths *et al.* 1983). Following the same symmetry, and breaking the glass very slowly, leads to good knives. The glass squares can be stored but it is advisable to use the knives immediately.

The major problem with this new approach to knife making is to get glass strips that have the required dimensions. The different knife making machines will have to be adjusted individually in order to break good knives, but it is certainly rewarding to invest some time to set up and align the knife-maker for optimal use (Stang 1988). It is generally acknowledged that the sectioning properties of even a perfect knife can be improved by a light tungsten coating (Roberts 1975; Griffiths *et al.* 1983). An apparatus for tungsten coating is presented in Fig. 3.5. The coating allows the production of thinner sections and the lifetime of the knife is greatly increased (the knives can be rinsed with distilled water and reused). Tungsten-coated knives can be stored for several weeks without losing their sharpness.

3.2.7 Freeze-sectioning and cryosectioning

Until recently it was relatively difficult to obtain good, ultrathin cryosections of biological material. This was partly because not enough was known about the treatment of the specimen prior to freezing and about the sectioning process itself (e.g. the effect of ice crystal size on sectioning, the sectioning temperature, etc.). Even though it is still too early to state that cryosectioning is a standardized routine procedure (although in quite a few laboratories it is at least a routine procedure), todays' cryosystems do facilitate cryosectioning to a considerable extent. Note, however, that a great many parameters—freezing method, use of cryoprotectants, sectioning temperature, etc.—have to be established for the individual experimental system and specimen before success can be achieved.

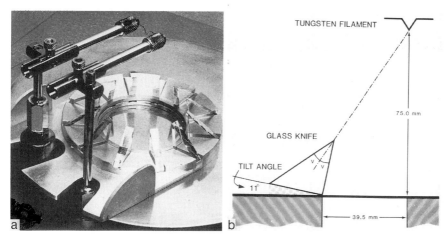

Fig. 3.5. Apparatus for coating glass knives with tungsten to improve their durability and sectioning performance; (a) photograph of the coating arrangement and jig; (b) schematic diagram to illustrate the geometric relationship between the tungsten source and the knife mounted in the jig. (Courtesy E. Stang.)

The process of cryosectioning is not yet completely understood, and is therefore the subject of many studies (Saubermann *et al.* 1977; Frederik *et al.* 1982; Karp *et al.* 1982; Chang *et al.* 1983) and heated arguments. Is it a real cutting process (deformation in a plastic material) or a fracturing process (in a solid)? Are cryosections cut by trough-melting at the knife edge? There seems to be general agreement that melting does occur, but that the extent of melting is insignificant (Karp *et al.* 1982; Chang *et al.* 1983). Estimations of the depth of the melting zone range from 0.4 nm to 20 nm and will be dependent on the nature of the specimen, the specimen temperature, and the knife temperature.

The most critical parameter affecting cryosectioning is the status of the specimen after freezing. If the freezing quality is good and ice crystals within the specimen are small or even absent (vitrified specimen), sectioning is greatly facilitated (Chang *et al.* 1983). Different applications require different sectioning temperatures, and the most suitable temperature is determined by trial and error. For example, for immunocytochemical purposes (see Chapter 6), sectioning temperatures around 210 K for light microscopical studies and around 180 K for electron microscopical studies will usually be sufficient, depending on the hardness of the block. The hardness of the block can be varied by varying the sucrose concentration in the infusion medium: the lower the sucrose concentration, the harder the block, and thus the lower the required sectioning temperature. If the specimen is not cryoprotected, and the sections are to be used for subsequent X-ray microanalysis, the sectioning temperature should be kept below 150 K. Morphological studies of fully-hydrated vitrified specimens require sectioning temperatures as low as 120 K (details in Chapter 5). Whatever temperature is found to be optimal for the given specimen, thermal conditions within and around the chamber should be stable whilst sectioning.

Cutting speeds are usually slow ($\leqslant 1$ mm s^{-1}) to prevent a high energy input into the specimen, and the knife clearance angle is of the order of 4°. Sectioning is preferably done on a dry knife. Wet sectioning onto a trough of liquid kept at low temperatures has been described (Hodson and Marshal 1970), but the general consensus is that cryosections should not make contact with liquids during the sectioning process. The sections will come off the knife quite compressed and will need to be manipulated with an eyelash probe. Eyelash probes can easily be made by glueing a fine stiff hair (those from short-haired dogs are found to be particularly good) to the end of a plastic straw or wooden stick.

Sections cannot be obtained with an accurately predetermined thickness; the thickness setting on the microtome is for general guidance only. The judgement of thickness using interference colours is relatively difficult. For some reason sections often charge electrostatically, and as a result section handling and mounting onto a grid can be a frustrating experience providing low yield for the effort expended. Antistatic devices can help to reduce the problem.

Under optimized conditions sections should come off the block as small, glossy, transparent, cellophane-like pieces. If the original freezing was bad, the block containing large hexagonal ice crystals will be white and the sections will crumble during sectioning to produce 'snow'. Sections will be transferred to grids which are usually coated with a thin plastic/carbon film rendered hydrophilic by glow discharging. Hexagonal grids are recommended for immunocytochemistry and 50 mesh grids for X-ray microanalysis: the former for the rigidity they provide for the support film so that it withstands the labelling and staining protocol, the latter to enable sections to be mounted clear of grid bars. Subsequent handling and processing of the sections will differ according to the application (see Chapters 4, 5, and 6).

3.2.8 Problems and artefacts

As Professor H. Sitte put it in one of his lectures: 'Unfortunately nowadays it seems to be common usage to publish good results exclusively and to be bashfully quiet about all problems and mistakes, which are often connected with new methods and instrumentation'. To amplify this sentiment we would like to draw attention to artefacts in the cryosections introduced by the sectioning process itself and by subsequent section mishandling. Most of the artefacts described here can be found in sections of plastic-embedded material as well. For detailed descriptions see Chang *et al.* (1983), Dubochet and McDowall (1984*a,b*), and Dubochet *et al.* (1988).

3.2.8.1 Changes in the status of water in the sample

A specimen that was originally well frozen (vitrified), can suffer heat-induced transformations during sectioning (for example from amorphous to cubic or even hexagonal ice). Heat is generated by frictional forces between the knife and the specimen when sectioning and this can lead to beam sensitive bands, probably due to periodic cutting-induced devitrification. Melting and refreezing of a thin surface layer can cause the sections to appear grainy. The practical solution is to cut at temperatures that are as low as possible but still produce intact transparent sections, and to use low cutting speeds.

3.2.8.2 Chatter

Chatter occurs in both vitrified and crystallized specimens, as well as in plastic sections, and is caused by vibration of the specimen in relation to the knife. The result is that each section possesses a varying thickness across the axis perpendicular to the knife edge.

3.2.8.3 Deformation/compression

Sections are compressed in the direction perpendicular to the knife edge. Compression therefore reduces the dimension of any structure along the cutting direction by 30–50 per cent (Zierold 1982; Dubochet and McDowall, 1984a,b), and increases the thickness of the section by the same factor. Compression in thin sections is more pronounced than in thick ones. It is independent of the sectioning speed and temperature, and does not seem to be alleviated by the use of diamond knives. Compression cannot be avoided.

3.2.8.4 Crevasses

Crevasses are found in most non-vitrified sections and sometimes even in vitrified ones. They appear as a network of lines, and are fractures that go deep into the section, more or less parallel to the knife edge. They are probably formed by bending stresses on the side of the section which was inside the block before cutting (Frederik *et al.* 1982; Dubochet and McDowall 1984a).

3.2.8.5 Inadvertent freeze-drying

Freeze-drying of the sections, either intentionally for elemental analysis in the microscope or accidentally by careless handling, will affect high resolution structural observations, since ice crystal formation or growth will take place upon warming of the sample to the required drying temperature. Crystal formation destroys structures in the 10 nm range (Hall and Gupta 1983). In addition, linear shrinkage of 10–20 per cent of the original length in the fully-hydrated state can occur during freeze-drying (Zierold 1984).

...

3.2.8.6 Knife marks and other less severe artefacts

Sections are sometimes contaminated with 'snow' or hexagonal ice crystals and show knife marks. These marks are more pronounced when using glass knives than when using diamond knives, and they form on the side of the section which was the surface of the block before cutting.

Most of the artefacts described here, and partly summarized in Fig. 3.6, cannot be avoided. However, they can be reduced by monitoring the sectioning parameters carefully. Artefacts introduced before sectioning are usually attributable to freezing damage, the extent of which is dependent on pre-treatment and on the nature and composition of the specimen.

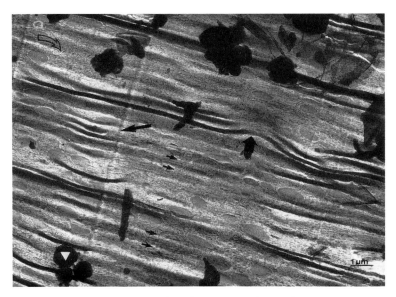

Fig. 3.6. Micrograph of an ultrathin fully hydrated section depicting artefacts associated with cryosectioning, such as crevasses (empty arrow), chatter (small arrows), compression (bold arrowhead), knife marks (big arrow) and contamination with atmospheric hexagonal ice (white triangle). (Courtesy A. McDowall.)

3.3 Summary: guidelines to successful cryoultramicrotomy

For most practical purposes, cryoultramicrotomy is best considered to be ultra-microtomy performed at low temperature. Good microtomy practices will result in successful cryoultramicrotomy, as long as the specimen has been well frozen. The key to the preservation of biological structure and composition is good freezing, preferably vitrification (see Chapter 2). The following factors will determine the success of the sectioning procedure:

1. Bumping the microtome support will result in chatter.

2. Resting forearms and elbows on the cryoattachment or other microtome parts during sectioning will result in chatter.

3. Large temperature fluctuations in the room (caused by draughts, breezes, or ventilation systems) lead to inconsistent sectioning. The specimen-surface temperature and the knife-edge temperature together determine whether or not the frozen specimen can be successfully sectioned in the automatic mode—this of course being the preferred mode if extraneous vibrational interferences are to be minimized. Specimen and knife temperatures are critically affected by the microtome chamber temperature. Any disturbance of the chamber temperature by external fluxes leads to inconsistencies. For example a temperature rise of 1 K would cause a 50 nm increase in the length of a one-inch block of aluminium, i.e. an increase in section thickness from 100 nm to 150 nm. In addition to eliminating draughts from the working environment, it is, therefore, also advisable to keep the microtome arm in continuous motion in automatic mode, so that thermal stability within the chamber is maintained during a sectioning session.

4. When sectioning plastic-embedded specimens, one good indicator of a sound sectioning arrangement is a well-polished block face. If the plastic face is not well polished, the chances are that the sectioning is not proceeding smoothly. There are several possible sources of difficulty:

(a) The plastic block may not be well polymerized. The analogy in cryo-ultramicrotomy is that the frozen block may be poorly frozen, resulting in 'snow' coming off onto the knife. If the block face is not polished, trying to section is a waste of time. A well-frozen block sections over a wide temperature range.

(b) The specimen may be loose.

(c) The knife may be loose.

(d) The knife may be dull.

(e) The clearance angle may be inappropriate, and may require adjustment.

5. All the tools should be clean and dry before cooling in liquid nitrogen.

6. All the tools used for specimen handling (directly or indirectly) or for tightening the specimen holder, tightening the knife, adjusting the knife, etc., have to be cooled by immersing in liquid nitrogen, or by continuous storage in the cryochamber of the microtome. This will prevent unnecessary, and possibly destructive, temperature rises in the specimen and help to maintain temperature stability within the cryochamber.

7. Frost or specimen fragments accumulated on the knife edge can be washed away with liquid nitrogen.

8. Finally, some of the artefacts described can be reduced with the

routine usage of good quality tungsten-coated knives. However, good cryosections can be, and have been, cut on conventional glass knives.

3.4 Conclusions

This chapter describes how to prepare ultrathin, fully-hydrated sections of unfixed, quench-frozen biological specimens containing at least some (peripheral) regions which are relatively undamaged by ice-crystallization. The cryomicrotomy of fixed specimens for histochemistry and immunocyto-chemistry is essentially similar, although section retrieval and post-sectioning treatments are quite different (see Chapter 6). Fully-hydrated sections of frozen tissues may now undergo further cryo-preparative manipulations so that they may be morphologically and/or analytically examined in either the hydrated or frozen-dried state in the electron microscope.

4 The transfer of cryosections to the TEM for morphological examination and X-ray microanalysis

4.1 Introduction

This chapter deals with the problems of physically transferring frozen sections to the transmission electron microscope in a state suitable for their eventual structural or analytical examination. It is worth repeating that (a) frozen sections are often full of artefacts, some avoidable others not, and (b) well-prepared frozen sections contain subcellular structures that are immediately recognizable, whilst the morphology of other structures appears somewhat different from that seen in conventional chemically fixed, resin-embedded preparations. For example, mitochondria possess a very dense matrix in unfixed, freeze-dried cryosections (Fig. 4.1).

Ultrathin frozen-hydrated sections mounted on filmed grids, and residing in the cryomicrotome chamber (see Chapter 3), are the important starting points of the discussion to follow.

4.2 Cold stages and transfer devices

The main problems associated with the morphological and analytical examination of frozen-hydrated sections are connected with specimen preparation (i.e. proper freezing, sectioning, section retrieval) and imaging (i.e. image formation and interpretation, avoidance of radiation damage). However, the structure and composition of the specimen can be seriously compromised at any stage during transfer from the microtome to the microscope.

Frozen-hydrated bulk specimens (see Marshall 1988) can be transferred to and maintained in the scanning electron microscope (SEM) with relative ease, because they can be mounted on a massive liquid N_2- or liquid He-cooled metal block, whose high thermal capacity and conductivity ensures that the specimen temperature remains extremely low, so that inadvertent freeze-drying is precluded. Furthermore, the large volume of the SEM specimen chamber, coupled with relatively undemanding imaging requirements, has meant that SEM cold stages and transfer devices have been

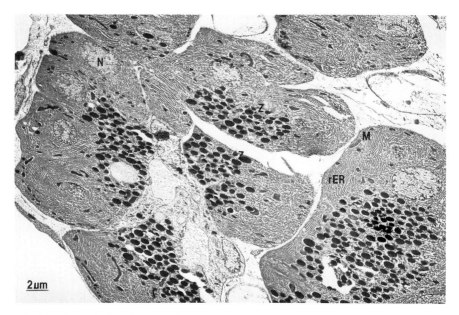

Fig. 4.1. Transmission electron micrograph of a thin frozen-dried cryo-section of quench-frozen rat exocrine pancreas; M = mitochondria, N = nucleus, rER = rough endoplasmic reticulum, Z = zymogen granules.

available and in fairly common use for some considerable time (Robards and Sleytr 1985).

Thin frozen-hydrated sections and aqueous films present immense technical difficulties, mainly because they are often mounted on support films whose thermal properties are less than favourable. Thus, the thermal contact between the specimen and the cold stage is indirect, which makes good contact between the edge of the grid or supporting annulus and the stage extremely important. If permissible within the context of the study being undertaken, coating the specimen with a thin layer of carbon (Zierold 1988) or aluminium improves its thermal stability. The confined specimen-chamber volume of the TEM, coupled with the TEM's high resolution capability, imposes formidable technical demands on the fabrication of a working cold stage.

A number of authors have published TEM cold-stage designs that permit high resolution imaging over a wide temperature range, and also reduce secondary radiation damage in organic specimens (e.g. Hayward and Glaeser 1980; Nicholson *et al.* 1982; Perlov *et al.* 1983). For those who wish to either build or modify their own cold stages and cryo-transfer devices, the cryo-engineering advice offered in the seminal publication by Heide (1982) is highly recommended. Fortunately, however, there are several excellent working systems available from commercial sources which are compatible with most modern TEMs.

The essential considerations in cold stages and cryo-transfer design are:

1. The cold stage must not compromise the optical performance of the microscope, and in particular the attainable resolution must not be reduced.

2. The cooled specimen must not drift due to mechanical instabilities and thermal contact with other microscope parts. In particular the cryogen reservoir and delivery system should not create vibrations.

3. The cold stage must be coupled with an efficient anticontaminator (or 'cold finger') so that there is no condensation of residual gases from the microscope column on the cooled specimen.

4. It must be possible to easily, rapidly, and accurately change the temperature of the stage within a wide range—this is important, for example, for facilitating the complete internal freeze-drying of cryosections prior to X-ray microanalysis (Zierold 1988).

5. It should be possible to make precise and continuous temperature measurements from near the specimen. (Of course, this measurement is not the true specimen temperature). See Heide (1982) for a discussion of the factors that affect the specimen temperature in the microscope.

6. It should be possible to tilt the specimen in the normal range.

7. There are two general types of cold stages: the 'top-entry' and the 'side-entry'. Efficient mechanically- and thermally-stable versions of both types are commercially available. Choice may depend on the specific application and the microscope to be used. Side-entry stages are by far the commonest, because they are compatible with the full functional range of modern goniometers. They are also more suitable for X-ray microanalysis. For example, it is possible to bring the solid-state detector very close to the specimen so that the solid-angle of X-ray collection is maximized and stray X-rays are minimized. In addition, the mass of metal in the line-of-sight between specimen and detector can be reduced to prevent X-ray 'shadowing' and to lower extraneous background contributions.

8. The specimen must be shielded by an efficient cryogen-cooled device (usually a spring-loaded shutter) during insertion into the microscope, so that a condensation layer does not form on its surface (Frederik and Busing 1986).

9. Mounting the hydrated specimen in the cold-holder should be easy, bearing in mind that (a) there should be little chance of affecting the state of the solid water in the specimen, and (b) thermal contact between the specimen support and the stage should be good.

An example of a commercially-available cold stage and transfer device is illustrated schematically in Fig. 4.2.

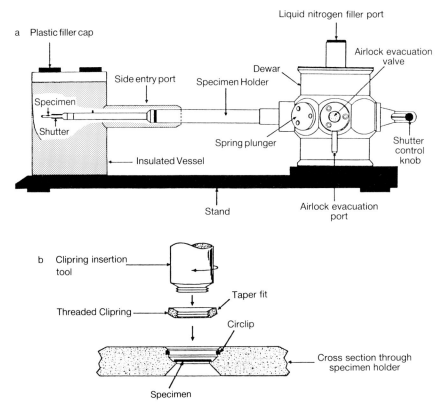

Fig. 4.2. Schematic diagrams of (a) the Gatan cold-stage parked in its workstation with the spring-loaded, liquid N$_2$-cooled, metal shutter withdrawn so that the stage is ready to receive a specimen mounted on a grid; (b) detail of the Gatan specimen holder showing how the grid is retained, and good thermal contact is established between it and the cold surface of the holder. (Reproduced with permission of Gatan Inc.)

4.3 Anticontamination devices

The vacuum of the electron microscope is not perfectly clean. If a cooled surface—for example an unprotected specimen mounted on a cold stage—is inserted into the vacuum, a condensation layer consisting mainly of water, hydrocarbons, nitrogen, and oxygen will form on it. At a pressure of 2×10^{-6} mbar a monolayer of condensate will form in about one second (Heide 1982). Indeed a very effective way of forming an amorphous ice layer is to insert an e.m. grid with a support film, mounted on a cold stage, into a microscope without the anticontaminator running.

Cooled specimens must, therefore, be shielded so that contaminants condense before reaching the specimen. There are several so-called anti-contaminators available; the features they should all share are:

1. They should be mounted close to the specimen, but should be constructed so that imaging and/or X-ray analysis functions are not impeded.
2. They should be cooled to a temperature below that of the specimen. (NB Some of the observed inconsistencies in the radiation resistance of cooled organic specimens were attributed by Cantino *et al.* (1986) to variations in the relative temperatures of specimen and anticontaminator).
3. Cold stage work should be performed in a vacuum that is as clean as possible to optimize the efficiency of the anticontaminator.

4.4 Freeze-drying of ultrathin cryosections

Frozen-dried cryosections of unfixed specimens show drying artefacts and are, therefore, of limited value for high resolution morphological studies despite their enhanced intrinsic contrast compared to hydrated sections (Dubochet and McDowall 1984*a*). They do, however, offer the possibility of analysing diffusible substances that otherwise would be lost or redistributed during chemical fixation (electrolytes), dehydration (electrolytes and lipids), and resin embedding (membrane components). Freeze-dried cryosections are consequently used for X-ray microanalytical studies (Hagler and Buja 1984; Somlyo *et al.* 1985; Roos and Barnard 1986; Warley 1986; von Zglinicki *et al.* 1987), autoradiography (Baker and Appleton 1976), and ion microscopy.

Freeze-drying of thin cryosections can be achieved at atmospheric pressure (Appleton 1978; Warley 1986) by leaving the hydrated sections on coated grids in the cold nitrogen environment of the cryomicrotome chamber. A more rigorous approach is to freeze-dry 'externally', i.e. in a vacuum unit other than the electron microscope column (Geymayer, *et al.* 1978; Barnard 1982; Somlyo *et al.* 1985; Wendt-Gallitelli and Wolburg 1984). In this procedure, the fully-hydrated sections are flattened and stabilized by a plastic film mounted over metal rings, and then transferred from the cryomicrotome in a liquid N_2-cooled lidded transfer device (Fig. 4.3) to a vacuum chamber. It is often convenient to use a conventional carbon-coating unit for this purpose. Under vacuum the liquid N_2 evaporates from the transfer device, and the sections warm up slowly and dry by sublimation before the water they originally contained can thaw. The vacuum is then released by purging with dry nitrogen gas, the lid of the transfer device is quickly removed, and the sections are coated with either carbon or aluminium. Sections are stored in a desiccator (with silica gel and not phosphorus pentoxide as the desiccant, because of the possibility of contamination) until required. The advantage of this technique is that it is cheap and simple, and it can cope with a number of different grids simultaneously. It is important to ensure that the specimen has reached room temperature, or has been warmed to a temperature slightly above ambient so

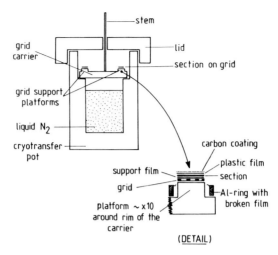

Fig. 4.3. Schematic section through a typical 'home-made' transfer device, made out of aluminium (for convenience), that can be used in conjunction with the freeze-drying of cryosections in, for example, a vacuum-coating unit. N.B. The device is fairly small (6–8 cm high), and should be parked in the cryomicrotome chamber; the external surfaces of the pot and lid act as cold fingers to prevent the condensation of atmospheric moisture on the sections. The grids should be handled with cold instruments, and all metal surfaces including tools should be clean and dry before usage.

that water does not condense on it during brief exposure to the atmosphere prior to coating.

To minimize the possibility of thawing and rehydration artefacts, it has been recommended (Hagler and Buja 1986) that the hydrated sections be transferred to the microscope in a liquid N_2-cooled cryotransfer specimen holder. Fortunately, three excellent side-entry holders and accessories are commercially available from the Gatan, Philips and Hexland companies— these are 'side-entry' holders which are compatible with the goniometer stages of most transmission electron microscopes (Fig. 4.2). Top-entry holders are available from Zeiss for their own TEMs. The hydrated sections can then be freeze-dried under controlled conditions in the microscope vacuum ('internal freeze-drying') by warming the stage gradually to temperatures in the region of 183 K to 188 K. A typical *cryo-chain* of activities from section retrieval in the cryomicrotome to viewing in the microscope is illustrated in Fig. 4.4. The disadvantages of internal freeze-drying, where the microscope is effectively used as a very elaborate freeze-drying chamber, are that only one section can be dried at any one time (although hydrated sections can be stored successfully under liquid N_2 for subsequent transfer and drying), and the microscope is committed until the viewing and analysis of that section has been completed. Amongst the major advantages of the direct cryotransfer approach is the ability to view and analyse a single section in both the hydrated and dehydrated states. It also

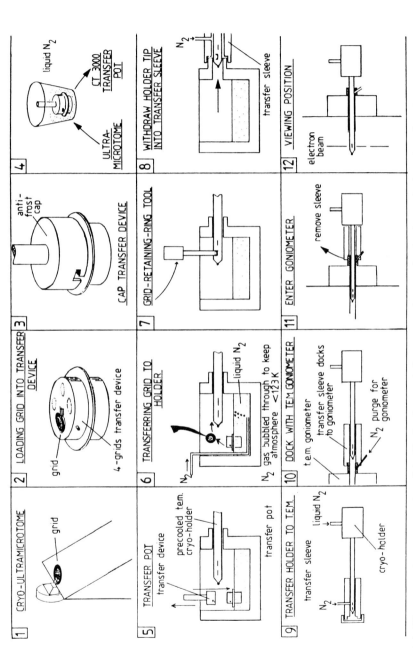

Fig. 4.4. Diagrammatic representation of the Hexland cryo-chain sequence for transferring frozen-hydrated sections from a cryomicrotome, via a grid-loading transfer pot, to the transmission electron microscope. (Figure re-drawn from an original kindly supplied by Hexland Ltd., now a member of Oxford Instruments plc.)

permits a definitive description of the state of frozen water in a section at different temperatures by electron diffraction.

4.5 Beam damage

The electron beam can affect irradiated specimens in a number of ways. It can (a) disrupt high resolution structural details (i.e. structural features in the range 0.5–1 nm, (b) disrupt crystal structure (discerned by the fading of electron diffraction patterns), (c) alter energy loss and optical spectra, and (d) remove organic mass and volatile elements, such as chlorine and sulphur (both aspects imposing serious constraints in quantitative X-ray micro-analysis). 'Radiation damage' is the term often used generically to describe this spectrum of disruptive events, and it is caused primarily by inelastic scattering. Organic biological specimens are especially prone to radiation damage. For further discussion of this important topic see Chapter 5.

4.6 Measuring the water content of sections

Although electron probe X-ray microanalysis is usually considered to be a physical method for measuring the cellular and subcellular distributions and concentrations of elements (Morgan 1985), the technique can also be used to measure the water content of cellular domains. Ingram and Ingram (1986) developed an indirect means of measuring subcellular water fractions by the analysis of freeze-dried sections infiltrated with a brominated resin, the rationale being that the embedding medium fills the space occupied *in vivo* by free water. A number of more direct methods have been introduced for measuring cell water in both hydrated and dehydrated cryosections. Some of these are summarized briefly here because of their considerable applied interest in physiology and pathology.

4.6.1 Continuum measurements on hydrated sections
        (Hall and Gupta 1979 and 1982)

In principle this is a simple method, based on the reasonable assumption that for biological soft tissues *the intensity of the continuum X-ray signal is proportional to the local mass-thickness* within the probed area. Thus, a local dry-weight fraction (F_d) can be determined from the ratio of the continuum signal in a given specimen compartment (a) whilst in its fully hydrated state on a cold stage, and (b) after freeze-drying in the microscope vacuum, by warming the cold stage to about 180 K:

$$F_d = \frac{(W - W_b)_d}{(W - W_b)_h} \tag{4.1}$$

where $(W - W_b)_h$ is the continuum X-ray signal (W) in the hydrated state,

(NB subscript 'h' = hydrated), corrected for the contribution to the continuum from extraneous sources (W_b), and $(W - W_b)_d$ is the corrected continuum in the dehydrated state (NB subscript 'd' = dehydrated).

Hall and Gupta (1979) found that eqn 4.1 is unreliable because of shrinkage during dehydration; this shrinkage tends to increase the mass thickness (mass per unit area), and hence the intensity of the continuum signal. Shrinkage can be corrected because elemental and continuum signals will be affected in the same way. Thus a version of eqn 4.1 that corrects for shrinkage is:

$$F_d = \frac{(W - W_b)_d}{(W - W_b)_h} \times \frac{(I_x)_h}{(I_x)_d} \tag{4.2}$$

where $(I_x)_h$ and $(I_x)_d$ are the characteristic intensities for any convenient element, x, before and after dehydration.

Equation 4.2 does not account for the fact that G values (where $G = \overline{Z^2/A}$ for the probed area, with Z the atomic number and A the mass number), and hence the continuum signal per unit mass thickness, are different in the hydrated and dehydrated states. A more accurate formulation is therefore advisable:

$$F_d = \left[\frac{(I_x)_h}{(I_x)_d} \right] \div \left[\frac{(I_x)_h}{(I_x)_d} + \frac{G_d}{G_{H_2O}} \left(\frac{(W - W_b)_h}{(W - W_b)_d} - \frac{(I_x)_h}{(I_x)_d} \right) \right] \tag{4.3}$$

Whilst eqn 4.3 is more accurate than eqn 4.2, in practice they both give similar results. For example, Hall and Gupta (1979) showed that if $(W - W_b)_h/(W - W_b)_d$ is 0.20, $(I_x)_h/(I_x)_d$ is 1, and (G_d/G_{H_2O}) is 3.2/3.67, then eqn 4.2 gives $F_d = 0.20$ and eqn 4.3 gives $F_d = 0.22$.

The water fraction F_w is given simply by:

$$F_w = (1 - F_d) \tag{4.4}$$

This approach to the measurement of cell water fractions possesses several important features:

1. It is not necessary to assume that the analysed section is uniformly thick.

2. The method depends on the ability to image and analyse sections in the frozen-hydrated state. Analytical spatial resolution is confined to the largest intracellular domains because of the radiation sensitivity of hydrated specimens (see Chapter 5).

3. Accurate corrections must be made to remove extraneous contributions to the continuum signal, a requirement which is especially stringent in the case of ultrathin sections compared to the 1–2 μm thick hydrated sections favoured by Hall and Gupta. (Whether it is possible to analyse ultrathin frozen hydrated sections by measuring the continuum signal under high electron dose conditions is a moot point, discussed in Chapter 5).

4. In practice it is considered preferable to make measurements before and after dehydration on adjacent fields, rather than on identical fields, because the original field is susceptible to severe latent damage on reirradiation after drying (Hall and Gupta 1979).

4.6.2 Continuum measurements on frozen-dried specimens with a peripheral standard (Rick *et al.* 1979)

In an analytical context, frozen-dried cryosections offer a number of advantages over frozen-hydrated sections: they possess higher intrinsic contrast, thus facilitating better *analytical* spatial resolution because finer structural components can be identified in real time (as opposed to in the darkroom); they are less prone to beam damage; and they may be handled with less elaborate and expensive ancillary equipment than frozen-hydrated sections necessitate.

The 'Rick method' (Rick *et al.* 1979) for measuring subcellular water was, paradoxically, developed for quantification in freeze-dried cryosections surrounded by a peripheral standard. Important features of the method are:

1. It requires a peripheral standard of known water content to be cryo-sectioned with the soft tissue specimen it encapsulates.

2. Dry mass, not water, is measured. By measuring the relative intensities of the continuum X-rays in the freeze-dried tissue compartments of interest and in the adjacent standard, water fractions can be determined.

3. It is necessary to assume that (a) the thicknesses of the section and its peripheral standard are equal and uniform at the time of sectioning (i.e. in the fully hydrated state), and (b) the specimen and standard experience the same degree both of lateral shrinkage during drying and of mass loss during imaging/analysis.

4. The mean atomic numbers (i.e. $G = \overline{Z^2/A}$) of the sample and standard are assumed to be the same.

In hydrated tissues, the dry weight fraction (F_d) of the specimen (spec) is defined as the dry mass (M_d) divided by total mass (M_t, i.e. M_d plus the mass of water). Therefore, in the specimen,

$$(F_d)_{spec} = \frac{(M_d)_{spec}}{(M_d + M_{H_2O})_{spec}} = \frac{(M_d)_{spec}}{(M_t)_{spec}} \tag{4.5}$$

(Note the similarities between eqn 4.5 and eqn 4.1.)

Fractional water content (F_w) can then be derived by simply substituting $(F_d)_{spec}$ in eqn 4.4. F_d can be determined by comparison with the peripheral standard (std) from the following general equation:

$$(F_d)_{spec} = \frac{(M_d/M_t)_{std}}{(M_d + M_t)_{spec}} = \frac{(M_d)_{std} \times (M_t)_{std}}{(M_d)_{spec} \times (M_t)_{spec}} \tag{4.6}$$

By assuming that in the hydrated state the specimen and standard are equally thick (thickness = mass/ρ, where ρ is the density), then dividing the numerator and the denominator in eqn 4.6 by (M_t/ρ_t) we have:

$$\frac{(F_d)_{spec}}{(F_d)_{std}} = \frac{(M_d)_{spec} \times (\rho_t)_{std}}{(M_d)_{std} \times (\rho_t)_{spec}} \tag{4.7}$$

But Hall and Gupta (1979) have demonstrated that $M_d = K \times (W - W_b)_d/G_d$, where K is a constant. Substituting this relationship into eqn 4.7 gives:

$$(F_d)_{spec} = \left[\frac{(W - W_b)_{dry\,spec} \times (G_d)_{std} \times (\rho_t)_{std}}{(W - W_b)_{dry\,std} \times (G_d)_{spec} \times (\rho_t)_{spec}} \right] \times (F_d)_{std} \tag{4.8}$$

This is the general expression of specimen dry weight fraction (F_d) that proves the basis of the Rick method. It assumes, as previously stated, that: (i) the specimen and standard experience the same degree of shrinkage, i.e. an identical volume is being analysed in the specimen and peripheral standard, (ii) the mean atomic numbers (G) of the specimen and standard are identical, (iii) the densities of the specimen and standard in the hydrated state are identical, i.e. $(\rho_t)_{std} = (\rho_t)_{spec}$. If the latter two assumptions are incorporated into eqn 4.8, we have an equation that can be used for the practical calculation of specimen dry weight fractions:

$$(F_d)_{spec} = \left[\frac{(W - W_b)_{dry\,spec}}{(W - W_b)_{dry\,std}} \right] \times (F_d)_{std} \tag{4.9}$$

4.6.3 *Water measurements in freeze-dried cryosections: a modified Rick method* (Warner 1986)

For most samples the Rick method (eqn 4.9), even with its intrinsic assumptions, appears to provide reasonable dry mass fraction values. This has been confirmed by comparative exercises on locust anterior caecum and toad urinary bladder (Hall and Gupta 1984). However, Warner (1986) recently drew attention to the shortcomings of the assumption that the densities (ρ_t) of the hydrated sample and standard are equal, and specifically to the inaccuracies that arise when the dry mass fractions of the standard and specimen are very different (e.g. if $\rho_{ice} = 0.93$ g/m^3, $\rho_{protein} = 1.32$ g/m^3, $(F_d)_{std} = 0.2$, and assuming $(F_d)_{spec} = 4 \times (F_d)_{std}$, then the error in calculating $(F_t)_{std}$ is 20 per cent and the error in calculating the water content is 40 per cent).

Warner (1986) offers an equation (presented here without its derivation) that reduces these errors and which may be useful for determining dry mass fractions in extremely heterogeneous specimens, where the mass and water content of the standard does not correspond closely with those of the specimen:

$$\frac{1}{(F_d)_{spec}} = \left[\frac{(W-W_b)_{std} \times (G_d)_{spec}}{(W-W_b)_{spec} \times (G_d)_{std}}\right] \times \left[\frac{1}{(F_d)_{std}} - \frac{(\rho_d - \rho_w)}{\rho_d}\right] + \left[\frac{(\rho_d + \rho_w)}{\rho_d}\right]$$

(4.10)

where all the attributes are of the specimen unless otherwise labelled.

The disadvantage of this method is that it requires knowing the density of the sample mass (ρ_d; ρ_w = density of water), although in practice it is encouraging that (a) the density of biological materials does not vary widely (e.g. $\rho_{protein}$ varies over a range 1.1–1.35 g/m^3), and (b) eqn 4.10 is not very sensitive to small change in density values (Warner 1986).

Caution must always be exercised when using peripheral standards to ensure that the artificial extracellular medium does not change intracellular composition.

4.6.4 *Water measurements in freeze-dried cryosections: mass estimation by scanning densitometry* (Zglinicki and Bimmler, 1987; Zglinicki *et al.* 1987)

This method makes use of the relative contrast levels of different cellular compartments measured in micrographs, and not the X-ray continuum signal, as an index of local specimen mass thickness. It is an interesting method because, due to the low dose conditions (low current, low magnification, low exposure time) under which the images are recorded, the specimen is much less likely to be subjected to the mass loss that inevitably accompanies the high-dose irradiation of biological specimens at ambient temperatures during the acquistion of X-ray spectra (see Chapter 5). The main features of this rather complicated method are:

1. It is assumed that (a) section thickness is constant over fairly short distances, notably in adjacent organelles, in fully hydrated cryosections, (b) there is negligible water translocation between compartments during specimen preparation, and in particular during freezing, and (c) there is no significant differential shrinkage during freezing.

2. Dry mass fractions in any given cellular compartments are estimated by combining, respectively, (a) the relative dry masses of cellular compartments measured by scanning microdensitometry, (b) the mean cellular dry mass fraction, measured either by computation from the dry mass fraction of the bulk tissue, or directly by X-ray microanalysis of bulk specimens initially in the frozen hydrated and then in the frozen-dried states, and (c) the volume densities of the cellular compartments estimated by stereological techniques.

4.6.5 *Other mass and water measuring methods*
(for reviews see Hall 1986; Zierold 1988)

Briefly, these methods include:

1. *Dark field signal measurement in STEM.* The basis of the method is that the dark field signal is linearly related to the mass thickness of thin specimens (Carlemalm *et al.* 1985*a*; Zierold 1986 and 1988). This method can be used to measure the dry mass of compartments in freeze-dried cryosections by making measurements on sectioned standards of known composition. An important practical requirement is that the thickness of the standard and cryosection must be identical.

2. *Beam attenuation.* One approach is to measure the attenuation of the total electron beam during passage through thin specimens (Halloran *et al.* 1978; Linders *et al.* 1982 and 1984). A conceptually related approach is to measure the attenuation of the zero loss peak with an energy loss (EELS) spectrometer (Hosoi *et al.* 1981; Leapman *et al.* 1984). These methods are complicated by the fact that electron transmission signals are exponentially, not linearly, related to specimen thickness.

3. *Backscattered electrons.* The intensity of this signal is linearly related to specimen thickness, and the method can be used both for sections much thicker than 1 μm and for ultrathin sections (Linders and Hagemann 1983; Linders *et al.* 1982).

4. *Densitometry.* See eqn 5.14.

For most biological purposes it would seem necessary to measure mass under low total electron dose conditions, where the amount of mass loss during irradiation is minimized (see Fig. 5.8 in Chapter 5). In the specific context of electron probe X-ray analysis, the initial low dose measurement of local mass can then be combined with the separate measurement, under high dose conditions, of characteristic elements (providing, of course, that the intense irradiation does not volatilize or disperse the heavier elements). An advantage of the alternative mass measuring methods mentioned above is that they do not require the often difficult (and critical) correction for the extraneous continuum.

4.7 Standards for quantitative X-ray microanalysis

For the quantitative analysis of frozen-hydrated cryosections it is common practice to prepare and handle, in exactly the same way as the specimen, cryosections of a salt solution in an organic matrix (e.g. gelatine, albumin, dextrose) mixture containing an added cryoprotectant (e.g. glycerol, dimethyl sulphoxide). These versatile standards can also be freeze-dried, again according to the same protocol as the specimen, and used for the quantitative X-ray microanalysis of freeze-dried cryosections (Hagler *et al.* 1983; Kendall *et al.* 1985). These are non-peripheral standards, i.e. they never come into contact with the specimen at any stage. A standard of this type which has all the essential analytical characteristics of cryosections, but is easier to produce, handle, and store is the aminoplastic standard. Amino-plastics are prepared by dissolving salt solutions in a self-polymerizing mixture of 50 per cent glutaraldehyde and 50 per cent urea (Roos and Barnard 1984; Roos and Morgan 1985).

Peripheral standards can be used in conjunction with both frozen-hydrated (Gupta *et al.* 1978) and frozen-dried (Rick *et al.* 1982) cryosections. These standards usually comprise a physiological saline made up in an organic matrix (e.g. dextran, albumin) in which the fresh specimen is immersed. The specimen is then removed from the solution, and the surrounding fluid layer is quickly removed until it forms a narrow (1 μm thick) 'capsule' around the specimen before quench freezing. It is essential to establish that peripheral standards do not interfere with the chemistry of the specimen.

Note that for most purposes absolute calibration is only required for one element; the concentration of other elements can, if necessary, be derived by the 'ratio method' from the relative sensitivities of the elements under given analytical conditions (Hall and Gupta 1983).

4.8 Summary: properties of thin cryosections

So far in this handbook we have seen how cryosections of native tissues can be produced by anhydrous methods (i.e. purely physical methods, without exposure to aqueous or organic media) and how they are transferred to the microscope ready for morphological and analytical examination. Before dis-cussing these features in detail in the following chapter it may be encouraging for the reader to be presented with a short preview of the properties of these specimens (Zierold 1984).

1. Freeze-dried cryosections must be protected from air. They can be rehydrated with ease by atmospheric moisture—with disastrous morpho-logical and analytical consequences. Contamination by suspended particles is also possible. For these reasons it may be advisable to freeze-dry hydrated cryosections under controlled conditions in the microscope vacuum, even though this produces a throughput bottleneck.

2. Ultrathin frozen-hydrated cryosections are especially susceptible to radiation damage. This is one of the reasons why Hall and Gupta (1979) sacrificed spatial resolution and performed their microprobe analyses on 1–2 μm thick cryosections. The minimum detectable concentration of elements is also reduced in thicker sections (Shuman *et al.* 1977).

3. Mass loss can be reduced by freeze-drying and by coating with a carbon film.

4. Frozen-hydrated sections of biological soft-tissues possess low intrinsic contrast, thus making 'real time' imaging difficult. Contrast can be enhanced by underfocusing.

5. Differential contrast in frozen-dried sections is good. However, linear shrinkage of 10–20 per cent occurs during drying.

In the near future it is anticipated that many of the imaging and mass loss problems that beset the 'conventional' continuum-normalization approach to X-ray microanalysis will be routinely overcome by the application of rapid on-line image analysis techniques, analogous to that described by Saubermann and Heymann (1987).

5 Imaging and analysis of frozen-hydrated specimens in the TEM: vitrified thin films and cryosections

5.1 Introduction

The cherished goal of biological electron microscopy is to be able to examine artefact-free preparations, in which the morphology and chemistry of the observed specimen bears a close resemblance to its *in vivo* state. Unfortunately, during most of the history of electron microscopy this ideal has not been achieved: in the case of fluid-suspended specimens, for example, damage is caused during 'conventional' preparative procedures by drying, adsorption onto the support film, and by contrasting/staining media (Adrian *et al.* 1984). The structural consequences of dehydration alone are potentially catastrophic.

Dramatic developments in the field of cryo-electron microscopy during recent years has changed the perspective. An early indication of the potential value of cryopreparations in conjunction with high resolution imaging was provided by Taylor and Glaeser (1974), who demonstrated by electron diffraction that 0.3 nm resolution could be preserved in frozen-hydrated catalase crystals. A further crucial development was the discovery that liquid water could, under certain circumstances, be cooled into a vitreous (i.e. glass-like) solid without the formation of structurally disruptive crystals (Brüggeller and Mayer 1980; Dubochet and McDowall 1981; Dubochet *et al.* 1983*a*).

The route was, therefore, signposted for the examination of certain favourable specimens 'infiltrated' and 'embedded', so to speak, in the most natural substance of all: non-crystalline water. This chapter will concentrate mainly on the methodology and high resolution TEM imaging of what may be described as 'vitrified thin film' specimens. Some general aspects of this exciting approach are (Dubochet *et al.* 1985):

1. In non-cryoprotected specimens, successful vitrification can only be achieved in thin aqueous films, preferably <1 μm thick. Specimen preparation is, however, easy and high yielding.
2. The advantages of cryo-electron microscopy are already apparent at

110 K, so that liquid nitrogen can be used as the coolant for microscope parts.

3. All the necessary hardware (cold stages and cold-transfer devices) is commercially available from a number of sources.

4. Whilst many workers in the field consider 'vitreous ice' to be synonymous with 'amorphous ice', it is not correct to use these terms indiscriminately and to consider amorphous ice as the juxtaposition of minute ice crystals not resolvable by direct imaging. [For a discussion of nomenclature see Newbury *et al.* (1986).] *The vitrified state is defined by the nature of the electron diffraction pattern.*

The electron microscopy of thin vitrified films can be subdivided into four major procedural steps:

(1) forming a thin layer of pure water, solute solution, or materials in aqueous suspension;

(2) rapidly cooling the films to the vitreous state;

(3) transferring to the microscope without re-warming above the devitrification temperature (~ 140 K);

(4) observing the specimen, in which the state of water has been characterized by electron diffraction, below 140 K with an electron dose low enough to preserve vitrified water.

In addition to describing these steps in some detail, we will discuss (a) how aldehyde-fixed, cryoprotected tissue specimens can also be vitrified, cryosectioned, and examined under low-dose conditions in the TEM without devitrification, and (b) how the high-dose conditions that are absolutely necessary for most electron probe X-ray analysis applications can result in severe radiation damage—an unwelcome effect that inevitably spawns serious quantitation constraints.

5.2 Preparing vitrified thin aqueous films

Despite the physicochemical arguments against the possibility of forming vitreous ice by rapidly cooling liquid water (Rasmussen 1982), it seems that vitrification of aqueous solutions and suspensions (of, for example, viruses and fibrous proteins) can be achieved in practice, but only if specimen size is reduced sufficiently to facilitate high cooling rates. Fortunately for the electron microscopist, specimens which are small enough to allow vitrification are most suitable for examination in the electron microscope. However, in forming thin aqueous films it is necessary to overcome the tendency of water to minimize the surface-to-volume ratio (a consequence of its high surface tension). Maintaining the fragile thin fluid layer long enough for it to be solidified by freezing is also a difficulty. Three simple methods have been developed for circumventing these obstacles (Dubochet *et al.* 1982*a*; Adrian *et al.* 1984; Dubochet and Lepault 1984).

5.2.1 'Thin film' freezing

Thin films (50–300 nm, but preferably < 200 nm) are produced on thin carbon films suspended over electron microscope grids (400 or 600 mesh). The key factor is the surface property of the support film. Support films normally tend to be hydrophobic and must be rendered *strongly hydrophilic.* This can be successfully achieved by glow discharging (Dubochet *et al.* 1982*a*; Lepault *et al.* 1983) for 10 s in an alkylamine atmosphere (e.g pentylamine or tripropylamine) at a pressure of 100 Pa, or in air (Namork and Johansen 1982). Dubochet *et al.* (1982*a*) suggest that droplets of stain solution sprayed at room temperature onto support films can test hydro-philicity; the films are too hydrophobic if the deposited droplets are circular with well-defined edges.

After glow discharging, a drop of aqueous solution dispensed from a micropipette is deposited on the hydrophilic film, with the grid held in fine forceps mounted on a simple guillotine-like frame above a liquid cryogen (liquid N_2-cooled ethane or propane) bath. Most of the fluid is removed from the drop by pressing briefly (~ 1 s, but a few trials will ensure fairly reproducible film thicknesses), between two pieces of dry filter paper (Whatman Qualitative No. 1). The grid is then plunged into the cryogen, and transferred to liquid N_2 using pre-cooled forceps. It may be stored under liquid N_2, or transferred carefully to the TEM cold stage (Fig. 4.2).

5.2.2 'Bare grid' method

Thin vitrified suspensions can also be prepared on hydrophilic grids without a support film. A deposited droplet is thinned by blotting, so that most of the grid squares are filled by aqueous films ~ 300 nm thick. The films can be thinned further by evaporation: the grid is observed under a low power microscope and when, after a few seconds, about 50 per cent of the occupied squares lose their films, it is plunged into liquid ethane.

A modification of the bare grid method, which yields more stable films, is the 'perforated film' variant. This consists of preparing thin vitrified films across the holes in a hydrophilic holey-carbon film (Adrian *et al.* 1984).

5.2.3 Spray freezing

With this method small microdroplets of solution are sprayed from a simple, commercially-available nebulizer on to a hydrophilic support film. In practice an electron microscope grid supporting the film is allowed to fall through the sprayed jet and into liquid cryogen; the delay between droplet deposition and freezing is short (<0.001 s) so that the liquid is not allowed to evaporate.

For most purposes the 'thin film' and 'bare grid' preparative methods provide adequate samples.

5.3 Preparing sections of vitrified bulk specimens

Frozen-hydrated sections of biological tissues are attractive for ultrastructural studies for many reasons. They preserve the structure and composition of subcellular and extracellular domains; they can be prepared and imaged at temperatures below the devitrification temperature and, therefore, damage caused by ice crystal growth at higher temperatures near to those necessary to effect complete freeze-drying is avoided; they are relatively resistant to electron beam damage compared to unfixed, unstained tissue sections observed at room temperature; they permit high resolution imaging—structures down to 10 nm can be resolved—and in general the structural features are similar to those seen in conventional plastic sections (McDowall *et al.* 1983). This impressive suite of claims can, however, only be realized if the original specimen is vitrified.

Crystallized water is almost impossible to cut into thin sections. In comparison, amorphous ice is considerably easier to section (Dubochet and McDowall 1984*a*), if not as easy as polymerized plastic. The problem is that the volume of fresh tissue specimen that can conveniently be sectioned in a cryoultramicrotome is much too great to allow complete vitrification. This problem can, in general, be avoided by pre-fixing 0.5–1.0 mm^3 blocks of tissue with glutaraldehyde, and then infusing them with a cryoprotectant such as sucrose (Dubochet and McDowall 1984*a*; Tokuyasu 1986*a*). It is possible to infuse glycerinated muscle with sucrose without aldehyde fixation (Fig. 5.1).

For ultrastructural studies the frozen blocks are sectioned at temperatures of around 113 K—well below the devitrification temperature—and transferred to the microscope cold stage for observation. These specimens are, in principle, embedded in endogenous water cooled to and maintained in the amorphous state.

5.4 Structure of ice: confirmation of vitrification

The structure of ice at normal or low pressure can take three forms (Dubochet *et al.* 1982*b*). *Hexagonal ice* (I$_h$) consists of large (μm) crystals, which, like all crystals, can be shown to contain defects such as grain boundaries, stacking faults, bend contours, and dislocations. *Cubic ice* (I$_c$) appears as a powdery mosaic of small crystals with average dimensions ~ 0.1 μm, that cannot be detected by electron diffraction when < 30 nm. *Vitreous*, or *amorphous ice* (I$_v$) is smooth and featureless, but possesses a number of

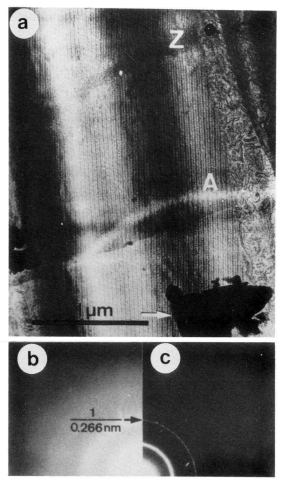

Fig. 5.1. Glycerinated flight muscle of *Lethocerus indicus* washed in 20 per cent buffered sucrose and frozen into the solid amorphous state by propulsion into liquid ethane. The sample was unfixed and unstained. Sectioning, transfer, and imaging were made at about 113 K with an accelerating voltage of 80 kV. (a) A nearly longitudinal section, thickness ~ 120 nm, Z- and M-bands are marked (Z and A, respectively). Arrows point to ice flakes deposited on the section. (b) Electron diffractogram of an amorphous solid specimen similar to the one shown in (a). (c) Electron diffractogram of the same specimen after it has been warmed to about 153 K. It corresponds to cubic ice; the second ring is located at a Bragg spacing of 0.266 nm. (With permission from Dubochet and McDowall 1984*a*.)

diffuse diffraction rings with well defined radii and widths. These three phases can be distinguished by their general appearance, and more definitively, by *electron diffraction* (Fig. 2.4). 'A material is (said to be) in a vitrified stage if, first, it is amorphous, and secondly, if it suffers a phase transition leading to a crystalline state when warmed above the devitrification temperature' (Chang *et al.* 1983).

Thin layers of pure water or of aqueous solutions can be frozen to the I_h, I_c, or I_v form depending on the thickness of the layer and the cryogen used. For example, I_v is obtained by spray freezing or thin film freezing of pure water layers < 1 μm thick when cooled by liquid ethane or propane; I_c crystals are formed with thicker layers (< 100 nm) frozen in less efficient cryogens such as melting nitrogen slush; pure water layers frozen in boiling liquid nitrogen always form I_h crystals (Dubochet *et al.* 1982*a*).

Three important practical points need to be underlined here. First, the form and the size of the ice within a specimen depends critically on the rate of cooling experienced by the specimen (Fig. 5.2). Second, the cryomicrotomist observing ice crystal 'ghosts' or 'imprints' in freeze-dried sections should refrain from loosely referring to the 'number' and 'size' of ice-crystals at different depths in the specimen. Much of the 'freezing damage' manifest in such specimens is caused by hexagonal ice (formed during initial freezing, and by transition and growth prior to freeze-drying). Individual I_h crystals can be very large, and multi-branched, extending over long tortuous distances into the specimen (Fig. 5.3). Electron diffraction of poorly frozen frozen-hydrated sections shows that much of the ice damage can in fact be attributable to a single, very large I_h crystal, or a small number of such crystals (Dubochet and McDowall 1984*b*). Third, it is possible to compromise the specimen severely after freezing because, under certain conditions, *phase transitions* occur from I_v to I_c, and from I_c to I_h (Fig. 5.4).

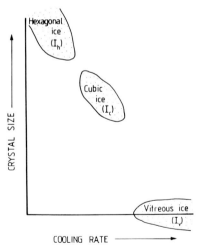

Fig. 5.2. Schematic representation of the type of ice and ice crystal size as a function of the rate of freezing. (Modified from Dubochet and McDowall (1984*b*).)

5.5 Imaging frozen hydrated specimens

Imaging frozen-hydrated specimens is difficult. The main problem is that

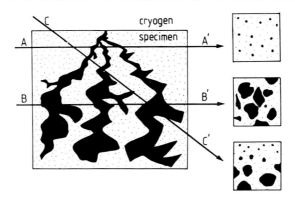

Fig. 5.3. Schematic diagram of a block of uncryoprotected frozen tissue (viewed in a plane perpendicular to the cryogen/specimen boundary) showing the presence of a single large hexagonal ice crystal that has nucleated near the specimen surface. (A), (B), and (C) refer to sectioning planes, and the appearance of 'ice crystal damage' or 'ghosts' vacated by ice during drying at different depths into the specimen. Cryosections would be easier to cut at (A), i.e. near and parallel to the surface where ice-damage is least serious. Sections would be difficult to cut several tens of μm into the specimen (B). Sections cut approximately perpendicular to the surface (C) would contain a relatively undamaged region.

frozen-hydrated specimens lack intrinsic contrast; in vitrified specimens, when it is considered that solute partitioning or segregation does not take place, the lack of contrast is due to small mass differences between aqueous and relatively non-acqueous compartments. This is one good reason why the majority of electron probe investigations have hitherto been performed on freeze-dried cryosections, where the contrast level is typically $3 \times$ better, but providing structural damage associated with the drying process (Chang *et al.* 1983) can be tolerated. However the imaging problem can be exaggerated: McDowall *et al.* (1983) remind us that the contrast level in unstained hydrated cryosections is superior to that in unstained plastic-embedded tissue sections. Indeed, the density difference between the specimen (1350 kg m^{-3} for protein) and the vitreous ice (930 kg m^{-3}) is considerable (Dubochet *et al.* 1985). In addition the signal-to-noise ratio is good; the 'noise' is low because there is no stain present, there may be no support film, and the specimen is relatively undamaged.

In the case of objects embedded in vitreous ice layers, Lepault *et al.* (1983) showed that it is possible to change their contrast by the addition of salts (e.g. sodium phosphotungstate, uranyl acetate, caesium chloride, the triiodinated glucose derivative Metrizamide) to the surrounding ice matrix. By adjusting the concentration of the 'stain' it is possible to obtain a range of contrast effects from standard negative staining through to positive staining. However, this interesting contrast embedding procedure does not represent a general solution to the imaging problem.

The consensus view is that for frozen-hydrated specimens, bright-field imaging in conventional transmission electron microscope (CTEM) mode

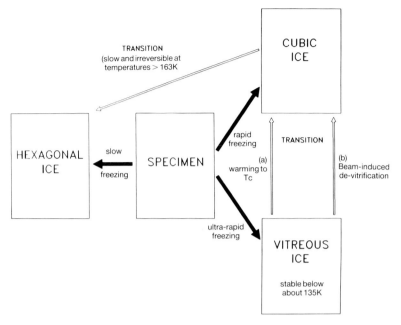

Fig. 5.4. Schematic representation of the relationships between the three major forms of ice. Under normal conditions the I_c to I_h transition cannot be observed *in the electron microscope*, because it takes place at a temperature (> 170 K) at which the evaporation of ice becomes significant. The I_v to I_c transition can occur in two ways when metastable I_v receives sufficient energy. First, by *warming* I_v layers up to about 135 K (i.e. T_c, the crystallization temperature); at T_c the direct image shows the sudden appearance of cubic crystals that remain unchanged on warming up to the point of significant evaporation, and the diffuse rings in the I_v diffracto-gram fade rather than getting sharper and are replaced by the sharp rings of I_c. Second, by *beam-induced* devitrification at a temperature between 110 K and T_c; the colder the specimen the higher the dose required to induce crystallization, until it becomes so high that the I_v layer is etched away before transition can occur. ICE IS NOT STATIC!

yields more information than the dark-field imaging generally used in scanning transmission (STEM) mode (McDowall *et al.* 1983). The reasons are that STEM is not well adapted for observing relatively thick specimens because of multiple scattering events, and it is less efficient than CTEM for observing small structures embedded in a thick layer. Contrast in the electron microscope is a mixture of amplitude contrast due to elastic scattering, and phase contrast due to inelastic scattering. Amplitude contrast dominates in the case of large objects, and phase contrast becomes increasingly important for smaller objects, until it becomes almost the only source of contrast for small organic objects less than about 6 nm in diameter. *To make the very best use of phase contrast, micrographs of frozen-hydrated specimens must be recorded with large defocusing values* (Fig. 5.5), because at Gaussian focus the image contrast is at a minimum. Enhancing phase contrast by underfocusing the objective lens results in a loss of resolution for certain

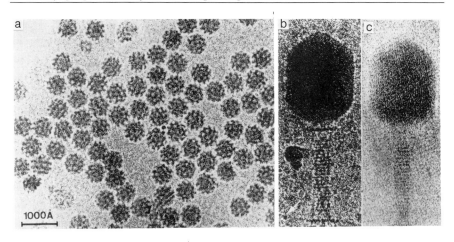

Fig. 5.5 The effect of underfocusing on the contrast level in frozen-hydrated specimens viewed under low-dose conditions. (a) Semliki Forest virus prepared by the perforated film method (electron optical magnification × 15 000; underfocusing 3.5 μm); (b) and (c) T4 bacteriophages in a vitrified film mounted on glow-discharged carbon films. The T4 tail structure is best seen in the more contrasting micrograph (b), where underfocusing = 2.5 μm; the head structure is best seen in (c), underfocusing = 0.5 μm. (With permission from Adrian *et al.* 1984.)

spatial frequencies (depending on the degree of defocus), so that some care must be exercised in the interpretation of fine details in strongly defocused images (Adrian *et al.* 1984). Accurately defocusing by a predetermined amount is an important prerequisite. Johansen (1973) and Meek (1976) describe how to calibrate an electron microscope so that defocus values are converted into decrements ('clicks' or 'steps') of objective lens current. In practice, the necessary calibration curve is normally provided in the microscope operators' manual.

5.6 Beam damage

Since it has become possible to examine thin biological materials in the fully hydrated state in the electron microscope, at least two important questions arise concerning the influence of electron irradiation on the specimen. First, can the frozen-hydrated specimen withstand irradiation under conditions that permit (a) high resolution imaging, and (b) electron probe X-ray microanalysis? Second, does the presence of water make the organic specimen more or less sensitive to beam damage? Before proceeding to describe some of the effects of electron fluxes on hydrated films maintained at low temperatures, it is worth noting that beam damage is reduced around liquid nitrogen temperatures (Glaeser 1975; Glaeser and Taylor 1978; Cantino *et al.* 1986). At very low exposures the 'cryoprotective' factor, as measured by the fading of the diffraction pattern in catalase crystals, is about × 3 (Lepault *et al.* 1983). Furthermore, it is disappointing (cost and operational complexities notwithstanding) that very little further radiation resistance is achieved by cooling to near liquid helium (< 10 K) temperatures (International Experimental Study Group 1986).

5.6.1 Low-dose imaging: general considerations

All organic specimens are potentially subject to damage during irradiation in the electron microscope. The definition of 'damage' will obviously depend on the purpose of a given investigation: subtle structural changes induced by relatively mild electron exposures may produce significant artefacts in the context of high resolution structural studies; the loss of organic mass occurring at higher exposures are of more concern to the microanalyst. To emphasize how alert the biological electron microscopist must be to the possible intrusion of radiation effects even if the observed specimen is mounted on an efficient cryostage, one can do no better than paraphrase the graphic example presented by Agar *et al.* (1974). A typical and comfortable current density on the viewing screen or at the negative for a dark-adapted TEM operator is about 3×10^{-11} A/cm^2. If the magnification is a modest $\times 10\,000$, the flux delivered in 1 second to the specimen is 3×10^{-3} A/cm^2 (i.e. a direct function of the square of the magnification) which is equal to 30 C/m^2 (since 1 Coulomb = 1 A/s). But since at 100 kV accelerating voltage, 1 C/m$^2 \equiv 40$ Mrad (note that this conversion is inversely related to energy, thus the relative damage suffered by the specimen decreases with increasing accelerating voltage (Reimer 1975)), the radiation dose suffered by the specimen in only 3 s would be about 4000 Mrad. Doses of this order of magnitude occur near the centre of nuclear explosions (Reimer 1975)!

Glaeser (1975) presented an expression that relates the contrast and resolution in an image to the electron exposure of the specimen, which demonstrates how the maximum dose that can be applied to a specimen without incurring severe damage depends on the size of the structure, i.e. radiation damage limits resolution.

$$(C \times d) > \frac{5}{\sqrt{f \times N_{Cr}}} \tag{5.1}$$

where C = object contrast (e.g. 0.1); d = minimum object size that can be seen ('resolved'); f = some 'utilization factor' which represents some preselected margin of safety in the exposure used (e.g. $\times 0.25$); the integer 5 refers to the necessity for the true or intrinsic object contrast to be about $\times 5$ greater than the point-to-point statistical fluctuations in contrast (i.e. background noise) if the object is to be seen with some certainty in the image (N.B. the signal:noise ratio in vitrified thin films is very favourable); N_{Cr} = critical exposure of damage, expressed as e$^-$/nm^2. (Note that since defining damage is somewhat subjective, a more quantitative assay of damage is given by the characteristic damage dose, $D_{1/e}$, which is the dose at which the intensity of a given signal falls to $_{1/e}$ of its original value at zero dose (Isaacson 1975). See, for example, the use of $D_{1/e}$ by Cantino *et al.* (1986) for defining the change in the continuum X-ray signal as an index of mass loss damage, and by Lepault *et al.* (1983), for defining the reduction in the intensity of the electron diffraction patterns of frozen-hydrated and freeze-dried catalase crystals).

According to this inequality ('eqn' 5.1), any combination of contrast and resolution that is less than the right hand side of the expression cannot be detected without increasing the exposure from the useful range and beyond the critical exposure point where the specimen is unacceptably damaged.

The electron dose required to produce a high resolution image is in the order of 100 e$^-$/nm^2 (Taylor 1978; Cosslett 1978). Thus, a subminimal dose or 'safe' dose in electron microscopy is considered to be near the lower end of the 5–100 e$^-$/nm^2 range (where 6.25 e$^-$/nm^2 = 1 C/m^2). (Note that the usual dose for aligning a column is about 62 000 e$^-$/nm^2 i.e. $\equiv 4 \times 10^5$ Mrad!). High resolution morphologists have long been aware of the constraints imposed by electron irradiation upon hydrated and dehydrated biological specimens, whether observed at ambient or low temperatures. Microprobe analysts have, with a few notable exceptions, been more cavalier in their attitude to specimen stability during electron exposure.

5.6.2 *Measuring electron exposure*

How, therefore, is the electron dose experienced by an object of interest during image recording or analysis (i.e. not including its exposure during preliminary 'searching' and focusing) measured? The most precise measurements, and probably the most convenient in the context of microprobe analysis, can be made directly with a Faraday cage. However, a method used widely in high resolution morphological studies makes use of the photographic exposure meter built into most TEMs. The optical density (D) of a photographic negative is defined as:

$$D = \log_{10}\left(\frac{I_0}{I}\right) \tag{5.2}$$

where I_0 = intensity of the incident radiation (i.e. light or electrons), and I = intensity of the transmitted radiation.

Thus, a region of the negative transmitting 10 per cent of the incident radiation has a D value of 1.0. It is important to appreciate that the D value of a photographic emulsion exposed to electrons (but not to light photons!) is directly proportional to exposure for most combinations of EM emulsions and developers up to D values of at least 1.5—this being the result of the 'single-hit' phenomenon, whereby each incident electron causes at least one silver halide grain in the emulsion to become developable. Therefore, for mild exposures:

$$D = k \times E \tag{5.3}$$

where E = the exposure (C/m^2 or e^-/nm^2), and k = a constant.

Because of this linear relationship, electron doses in the TEM can be estimated either from the optical density of the photographic film taken through a hole in the specimen at a given magnification and exposure time (Talmon *et al.* 1986), or from the microscope exposure meter which has been calibrated to an appropriate film exposed and developed under carefully standardized conditions. The absolute calibration is normally undertaken in both cases by the direct measurement of beam current collected in a Faraday cage.

The following three important points should be noted:

1. Every doubling of the image magnification results in a $4 \times$ increase in the exposure dose at the specimen plane.

2. Radiation damage is proportional to the total dose (i.e. the *total* number of electrons delivered per unit area of a specimen) and not to the rate at which it is delivered. Thus, a short exposure to a high intensity is as destructive as a prolonged exposure to a milder intensity.

3. Radiation damage is inversely proportional to the accelerating voltage (Glaeser 1975).

5.6.3 Mass thickness measurements

Mass thickness of thin films can be measured with some accuracy by densitometry. The method is based on optical density measurements made on micrographs with a microdensitometer, combined with the known scattering properties of the constituents of the specimen. The Heidelberg group (Dubochet *et al.* 1982*b*; Chang *et al.* 1983; McDowall *et al.* 1983) have successfully measured the mass thickness of frozen aqueous films and of cryosections by exploiting the relationship:

$$D = D_0 \exp(-W/\Lambda) \tag{5.4}$$

where D = the optical density of the image of a thin film or section; D_0 = the optical density of the photographic negative in the absence of the specimen (e.g. in a hole); W = mass thickness = ρt (where ρ is the density of the material being examined, and t is the thickness; it can be assumed that $\rho_{water} = 0.93$ g/m^3 (930 kg/m^3) and $\rho_{protein} = 1.32$ g/m^3 (1320 kg/m^3); Λ = mean-free mass thickness (expressed as g/m^2) of the material examined under a given accelerating voltage and objective aperture angle. Λ values can be read from Fig. 6 in Euseman *et al.* (1982). For example, Chang *et al.* (1983) found that under their microscope operating conditions that $\Lambda_{water} = 0.274$ g/m² and $\Lambda_{protein} = 0.281$ g/m².

Equation (5.4) transforms to give:

$$t = \left(\frac{\Lambda}{\rho}\right) \times \ln\left(\frac{D_0}{D}\right) \tag{5.5}$$

The data used in eqns (5.4) and (5.5) are very approximate, but they can yield a rapid estimation of specimen thickness, particle mass, the density of large particles, and the hydration of specimens by comparing the mass of the local area before and after freeze-drying (Chang *et al.* 1983; Dubochet *et al.* 1988). For a worked example see Euseman *et al.* (1982).

5.6.4 Practical low dose imaging

Low dose imaging in essence entails restricting the electron dose suffered by a specimen region of interest exclusively to that needed to expose the photographic plate, and therefore to eliminate dosing the chosen region during field selection, focusing, and other optical manoeuvres. Early 'blind shooting' procedures simply involved focusing on one region of the specimen and photographing another at relatively low magnification by moving the new region into the beam axis. The beam tilting procedures introduced by Williams and Fisher (1970) and Unwin and Henderson (1975) represented a major advance in high resolution microscopy of organic specimens. Beam tilting does not require the specimen stage to be moved to bring the region to be photographed into the beam axis, and it permits focusing and astigmatism correction to be performed at high magnifications (around × 100 000) on an off-axis region of the specimen which, nevertheless, lies immediately adjacent to the region of interest.

Wrigley *et al.* (1983) describe in detail an elaboration of the beam tilting procedure (Fig. 5.6). This method, requiring only minor modifications to a standard TEM column, enables the operator to obtain consistently any desired defocus value in photographs taken under low dose conditions and low magnification. It is not necessary to describe each of the several steps in the method but it is illustrative of the exacting requirements of low dose imaging to state that they fall into four sequential phases:

1. *Preliminary microscope adjustments.* In particular, the column must be accurately aligned.

2. *Search.* The region of the specimen to be photographed is located at *low magnification* (~ × 2000). (N.B. The dose delivered to the specimen increases as the square of the magnification.) Once the object of interest is located and centred the beam is immediately shuttered, both deflectors (i.e. above and below the specimen) switched on, then the shutter removed. The area adjacent to the object is now illuminated.

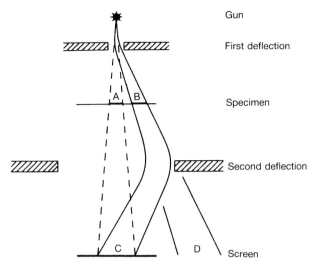

Fig. 5.6 Beam tilting scheme for use in conjunction with accurate defocus and low-dose imaging. It shows an axial beam (broken lines) illuminating an axial patch of specimen at A imaged at the screen centre C. With the first deflection switched on, the specimen is illuminated off-axis at B and imaged at D. With the second deflection also switched on the image at B is brought back to C. (N.B. This diagram is used here to illustrate the philosophy of modern low dose imaging; it is not intended that readers exclusively adopt this approach, excellent thought it is). (With permission from Wrigley *et al.* 1983.)

3. *Focus.* This is performed critically by observing the extinction of background granularity at very high magnification ($\sim \times 300\,000$). Defocus by a known amount to provide optimal phase contrast.

4. *Photography.* Reduce magnification to $\sim \times 20\,000$. N.B. Photography should be performed at the lowest tolerable magnification to ensure lowest possible electron dose.

A very useful protocol for the low dose imaging of frozen-hydrated thin films, that minimizes beam-induced specimen drift was provided by Dubochet *et al.* (1988). This involves: (i) selecting both the area to be recorded (R), and an area (S) at a relatively large distance from R to be used for focusing, at low magnification; (ii) focusing on area S with a preselected defocus value compatible with resolution requirements; (iii) switching to a preselected condenser position where the beam is spread to provide only sufficient illumination to enable the area R to be located; (iv) blanking the beam for a few seconds to allow R to stabilize; (v) switching the condenser to a second preselected position that provides the correct illumination for recording the image at a chosen magnification and exposure time.

An approach used to considerable effect for the high resolution imaging of radiation sensitive objects is *signal averaging*. This is applied most effectively to periodic structures, and entails acquiring several noisy images at low exposures well below the critical damage level, and then combining them in a computer to yield an acceptable image. Although the contrast observed in frozen-hydrated specimens is low, the very favourable signal-to-noise ratio more than compensates for this (Adrian *et al.* 1984), thus revealing high resolution structures in periodic and non-periodic specimens often without recourse to signal averaging.

Somlyo *et al.* (1977) adopted a '*spectral averaging*' approach, somewhat analogous to the morphological signal averaging, in an attempt to limit mass loss during the microprobe analysis of frozen-dried cryosections at 173 K. Although they reduced the total delivered dose by summing 4–6 spectra accumulated for 25 s at each different location, the dose values (~ 500 C/cm^2, i.e. $\sim 3 \times 10^7$ e$^-$/nm^2 in 50–100 nm probes; ~ 15 C/cm^2, i.e. $\sim 9 \times 10^5$ e/nm^2 in 4 μm probes) are still exceptionally high, and induce up to 15 per cent loss of specimen mass.

5.6.5 Radiation damage—its manifestations

Radiation damage, as observed by electron diffraction, mass loss and spectroscopic methods, is a function of *chemical structure* (e.g. compounds with benzene rings are more resistant than aliphatic compounds), *electron energy* (i.e. inversely related to accelerating voltage), *specimen temperature*, and the *total electron dose* (Glaeser 1975; Reimer 1975; Isaacson 1977). Dry biological material irradiated at room temperature is more susceptible to damage than frozen-hydrated specimens at low temperatures (McDowall *et al.* 1983), and damage phenomena are much more pronounced in hydrated specimens composed of crystalline ice than in vitreous ice (Talmon *et al.* 1986).

In the case of vitrified specimens, the I_v–I_c transition can be triggered by the beam between 110 K and T_c (see Fig. 5.4); the colder the specimen the higher the dose required (Dubochet *et al.* 1982*a*). Dubochet and Lepault (1984) identified a strange *beam-induced vitrification* phenomenon at temperatures below 70 K, i.e. I_h and I_c are transformed to amorphous ice in thin hydrated films irradiated with 200–500 e$^-$/nm^2.

'*Bubbling*' (Fig. 5.7) is an electron irradiation effect which is seen in pure water films, but is particularly noticeable in hydrated solutions of organic molecules (Dubochet *et al.* 1982*a*; Lepault *et al.* 1983; Talmon *et al.* 1986), and in frozen hydrated cryosections (McDowall *et al.* 1983). The phenomenon occurs in ice in the I_v state, and in the interstitial matrix between I_h crystals, but not in I_c ice. It appears after a dose in the 10^3–10^4 e$^-$/nm^2 range, although the dose required for bubbling increases with decreasing temperature. Furthermore, it is not strongly dependent on the concentration of the organic solution, and is only slightly dependent on the rate of electron delivery to the specimen (i.e. the flux, expressed as e$^-$/nm^2/s). It is hypothesized (Dubochet *et al.* 1982*a*; Talmon *et al.* 1986) that bubbling is the result of pressure generated by trapped molecular fragments formed during irradiation.

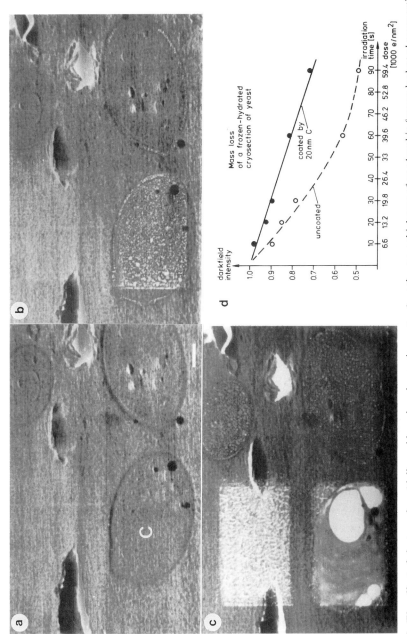

Fig. 5.7. The effect of electron dose (delivered in reduced scanning rasters) on the mass thickness of an ultrathin frozen-hydrated cryosection of yeast cells in suspension. (a) Undamaged, fully frozen-hydrated section showing a cell, C. (b) Exposure = 1300 e⁻/nm²; note (i) the extracellular growth medium is more radiation sensitive—the raster is more obviously luscent—than the cell matrix, and (ii) mass loss within the cell starts with the formation of bubbles within the outer membrane. (c) Exposure = 80 000 e⁻/nm². Bar = 1 μm. (d) Darkfield intensity, as a measure of the mass thickness of a frozen-hydrated ultrathin cryosection of yeast, in relation to the irradiation dose. Note the protective effect of the thin carbon coating. (With permission from Zierold 1988.)

Mass loss induced by relatively high dose electron irradiation is a problem characterized by the following features:

1. Mass loss is directly proportional to the electron dose (Fig. 5.7), but is independent of the thickness of the irradiated specimen. It is a surface etching phenomenon, not a local heating effect caused by scattering events. In frozen pure water films (I_v or I_c ice states) at 110 K, exposed to 100 kV electrons, about 60 electrons are required to remove one water molecule (Talmon *et al.* 1986; Dubochet *et al.* 1982*a*). Mass loss is not, therefore, a serious constraint in low dose, high resolution, morphological imaging of frozen-hydrated specimens.

2. The radiolytic mechanisms resulting in surface etching are poorly understood, although it has been suggested (Heide 1984; Talmon *et al.* 1986) that in hydrated specimens highly reactive free radical species are generated during the interaction of incident electrons and water:

$$e^- + H_2O \rightarrow e^- + H^\bullet + OH^\bullet$$

The $H^\bullet$ and $OH^\bullet$ radicals react with water and organic molecules to produce volatile fragments that escape from the surface of the specimen. Pure water films are more radiation resistant than aqueous films containing organic materials (Dubochet *et al.* 1982*a*) and frozen-hydrated tissue sections.

3. *Completely* freeze-dried cryosections may be about $\times 100$ more stable (i.e. can tolerate $\times 100$ higher dose before mass loss becomes apparent) than frozen-hydrated sections of the same material (Zierold 1988). Zierold (1988) did, however, emphasize the need to remove all the water by completing the drying process at temperatures in the 223–233 K range, otherwise the remaining water seems to promote severe mass loss by catalyzing the radiolytic processes described above.

4. Mass loss is reduced by lowering specimen temperature (Hall 1979; Shuman *et al.* 1976; Cantino *et al.* 1986). Under very favourable conditions the tolerable dose at 100 K may be $\times 10$ to $\times 100$ greater than at room temperature (Egerton 1980; Cantino *et al.* 1986), whilst a further stabilization factor of the order of $\times 2.5$ may be expected in cryosections irradiated at 10 K (Zierold 1988).

5. Mass loss presents an intractable problem in the quantitative electron probe X-ray analysis of biological materials, and especially of thin frozen-hydrated sections. The problem stems from several inescapable facts:

a. X-ray microanalysis can only be achieved by delivering very high electron doses to the specimen—it is necessary to deliver at least 20 nano-Coulombs (20 nC) to the analysed area to generate sufficient counts to meet statistical requirements (Hall and Gupta 1984; Zierold 1988). If we assume that the (hypothetical) analysed area is 1 μm^2, this is equivalent to a dose of about 10^5 to 10^6 e^-/nm^2.

b. Uncooled organic specimens suffer an extensive loss of mass at exposures of 10^3 e$^-$/nm^2 (i.e. 10^9 e$^-$/μm^2) to 10^4 e$^-$/nm^2 (i.e. 10^{10} e$^-$/μm^2) (Bahr *et al.* 1965; Cantino *et al.* 1986). This means that the *minimum* exposure necessary for analysis of a coarse 1 μm^2 area is about 1–2 orders of magnitude higher than the critical dose inducing significant mass loss! Of course the problem is even more serious when the analysis of small compartments is performed with finer probes.

c. Cooling the specimen to liquid nitrogen temperature on a cold stage improves the situation to a point where the analysis of frozen dried sections becomes just about tolerable with a 1 μm^2 probe. According to Hall and Gupta (1984) the tolerable dose at 100 K should be no more than 10^5 e$^-$/nm^2, but Zierold (1988) claims that freeze-dried cryosections are stable at 100 K up to electron exposures of 10^7 e$^-$/nm^2. These values cast serious concern on any quantitative microprobe based on the continuum method and X-ray analysis performed with probes much smaller than 1 μm^2, because the local dose experienced by the specimen, even if cooled to liquid nitrogen temperature, inevitably exceeds the dose that causes mass loss.

d. Since mass loss is predominantly a surface etching problem, specimen thickness must be an important consideration, especially in the case of radiation-sensitive hydrated specimens. Hall (1986) calculated from data for hexagonal ice (Dubochet *et al.* 1982*a*) that in a section of initial thickness 0.2 μm exposed to a relatively mild *analytical* electron dose of $\sim 3 \times 10^7$ e$^-$/ nm^2 in a probe of 0.01 μm^2 that the section would tend to be etched to a depth of 2 μm before the completion of analysis! Similar pessimistic calculations were presented by Cantino *et al.* (1986), Zierold (1988), and Shuman *et al.* (1977). What, therefore, are the practical implications of these findings? It would appear that it is feasible to analyse frozen-hydrated sections in the 1–2 μm thickness range (Hall and Gupta 1983, 1984) and bulk frozen-hydrated specimens (Marshall 1988) because the proportion of the total mass that is lost by surface sublimation is acceptably low. However, in thin frozen-hydrated sections the mass loss is prohibitively high.

In biological X-ray microanalysis the difficulties presented by mass loss are frequently side-stepped by assuming that the beam sensitivities of both the specimen and chosen standard are roughly equivalent, so that the effect cancels out. This assumption is unreasonable (Hall 1979) because it is well known that the mass loss end-points for various organic materials ranges from 10–90 per cent. Furthermore, it is not unlikely that the different cellular structures possess different radiation stabilities (Cantino *et al.* 1986). Whilst acknowledging that the prospects of quantitatively analysing, with the continuum normalization method, ultrathin frozen-hydrated and freeze-dried cryosections with spatial resolutions better than 0.1 μm are gloomy (Hall and Gupta 1984), it is worth considering how intrinsic problems can be minimized. Some of the solutions offered are: (a) to coat the sections with a thin conductive carbon film (Zierold 1988); (b) to analyse at or near liquid helium temperatures (Zierold 1988), although the poor cost:benefit ratio is unlikely to make this a routine practice in many laboratories; (c) to determine local specimen mass by independent low dose methods, e.g. by measuring electron transmission, the attenuation of the zero-loss peak in the electron energy-loss spectrum, and the intensity of backscattered electrons (for references and brief evaluation, see Chapter 4).

5.7 Conclusions

The vitrified thin film method (Adrian *et al.* 1984; Milligan *et al.* 1984) and the cryosectioning of vitrified specimens (Dubochet and McDowall 1984*a*) have been used to obtain new morphological descriptions of the high-resolution structure of viruses, membranes, ribosomes, and various organic macromolecules. The new approach can also challenge old beliefs. For example, the examination of frozen-hydrated preparations (Dubochet *et al.* 1983*b*) have shown that the bacterial mesosome is an artefact produced during conventional wet chemical e.m. preparation procedures. Thus, the scope of these elegant techniques is enormous. However, to achieve the impressive feat of imaging fully hydrated biological specimens without the intrusion of ice crystal and electron irradiation damage requires a sound multidisciplinary grasp of the principles of cryobiology, cryotechnology, and electron optics.

6 The use of ultrathin cryosections in immunocytochemistry

6.1 Introduction

Labelling of the cell surface and the intracellular components for microscopic observation is an important technique in studying their localization and function in the cell. A wide range of cytochemical and immunocytochemical methods has therefore been developed. Some of the methods are relatively unspecific (e.g cationic dyes, cationized ferritins). Others are very specific (e.g. autoradiography, colloidal gold-conjugated macromolecules). Specific labelling procedures can be divided into two classes: (a) the diffuse labels (autoradiography, fluorescent or peroxidase conjugates), and (b) the particulate labels (ferritin, colloidal gold).

One of the most versatile particulate markers is colloidal gold (Beesley 1989). It can be produced in a wide size range from 5 nm to 150 nm with particles of even size, and can be conjugated to a variety of different macromolecules, for example:

1. *Enzymes* such as RNAase and DNAase can be used for localizing specific substrates.

2. *Lectins*, which are proteins or glycoproteins with binding sites that recognize a specific sequence of sugar residues (e.g. concanavalin A with a sugar specificity for α-D-glucose and α-D-mannose), can be used for localizing sugar residues, glycoproteins and glycolipids (Leathem 1986).

3. *Streptavidin* is a protein isolated from the culture medium of *Streptomyces avidini* which possesses four sub-units, each with a high affinity biotin-binding site. Hormones, lectins, nucleic acids, and antibodies can be biotinylated and can thus be visualized by streptavidin-gold. Since antibody molecules can each be coated with up to 150 biotin molecules each of which can in turn bind to streptavidin, this assures a high label : antigen ratio (Varndell and Polak 1987).

4. *Protein A*, isolated from the cell wall of *Staphylococcus aureus*, is a protein with the ability to bind to immunoglobulins from several species, thus allowing the conjugate to be used with a large variety of different primary antisera (Horrisberger and Clerk 1985).

5. *Protein G*, isolated from the cell wall of a group G streptococcal strain G-148, is a protein with the ability to bind to immunoglobulins from several

species (Åkerström and Björck 1986) which seems to be a more versatile IgG-binding reagent in immunocytochemistry than protein A (Bendayan 1987).

6. *Antibodies*, which are special proteins generically called immunoglobulins (Ig), have the unique ability to recognize and bind to a single antigenic group. The predominent immunoglobulin produced in secondary antibody response is IgG (there are another four classes of antibodies in mammals, IgA, IgD, IgE, and IgM), which is thus the one most often used in immunocytochemical localization studies.

The coating of the gold particles is a non-covalent binding process which is possible because of their surface charge. Colloidal gold, a strong emitter of secondary electrons, is very electron-dense and small enough not to obscure fine cellular structures. It is therefore suitable for both SEM and TEM studies (Horrisberger and Rosset 1977; Horrisberger 1984) and can be used for multiple labelling. In this chapter we will describe the use of colloidal gold in connection with immuno-electron microscopy, in which it is coated with antibodies (Faulk and Taylor 1971; Romano *et al.* 1974), protein A (Romano and Romano 1977; Roth *et al.* 1978), or antigens (Larsson 1979).

Immunoreagents labelled with colloidal gold have been used successfully to (a) label the cell surface antigens of cells in fluid suspension and of cells grown as monolayers, and (b) detect intracellular antigens in ultrathin plastic sections or thawed cryosections. For electron microscopical studies the gold label does not need signal amplification, and this makes it a suitable candidate for quantitative studies. Some of the more common labelling methods are presented in Fig. 6.1. They are: the direct labelling (Fig. 6.1a), the indirect labelling (Fig. 6.1b), and the protein A–gold labelling (Fig. 6.1c). In addition, Bendayan (1982) developed a 'two-face' double labelling procedure that is limited to resin embedded sections mounted on uncoated grids, and can be used with either protein A–gold complexes or secondary antibody–gold complexes. The idea is to label different antigens that are co-localized with two different markers, one applied from one side of the section and the other from the opposite side. Another, not widely used, technique is described by Larsson (1979) where a surplus of primary IgG ensures that only one of the two antigen-combining sites of a given antibody reacts with the antigen. Colloidal gold coated with the antigen will then react with the second antigen-combining site.

It is not quite clear whether the protein A–gold or the secondary antibody–gold (i.e. IgG–gold) complexes give the better results, but the protein A–gold method has some distinct advantages. These are: (a) it can be applied to antibodies from many different species, and (b) it reacts with the primary antibody in a 1:1 ratio under saturation conditions, which is of importance for quantitative studies. The latter is a disadvantage if signal amplification is desired. Methods for the production of a colloidal gold solution with

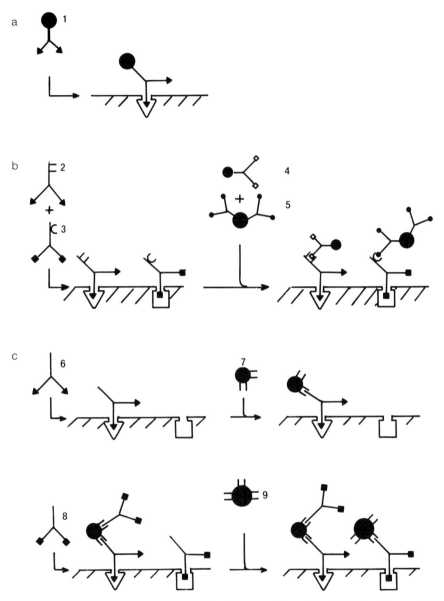

Fig. 6.1. The three most commonly used labelling procedures: (a) colloidal gold coated with primary antibody (1); (b) colloidal gold coated with secondary antibody (4, 5) which recognizes the primary antibody (2, 3); (c) colloidal gold of different sizes (7, 9) coated with protein A. Protein A recognizes the Fc part of IgG molecules from several species (e.g. 6, 8).

Table 6.1. *Preparation of colloidal gold (Slot and Geuze 1985)*

1. Prepare solution A by mixing 1 ml of 1% $HAuCl_4$ solution with 79 ml distilled water.

2. Prepare solution B (reducing mixture) by mixing 4 ml of 1% trisodium citrate. $2H_2O$ solution with a variable volume (0–5 ml)* of 1% tannic acid† solution and 0–5 ml of 25 mM K_2CO_3 solution (same volume as the tannic acid) to correct the pH of the reducing mixture. Add water to 20 ml. Omit K_2CO_3 if tannic acid volume is < 0.5 ml.

3. Heat both solutions on a hot plate to 333 K.

4. Add solution B quickly to solution A, while stirring.

5. After sol formation is complete‡ (colour turns dark red) heat solution to boiling.

---

*The smaller the volume of tannic acid the bigger the gold particles (e.g. 3.0 ml tannic acid yields 3.5 nm gold particles and 0.5 ml tannic acid yields 6 nm gold particles).
†Aleppo tannin, Mallinckrodt, St. Louis, Code 8835.
‡Completion of sol formation is dependent on tannic acid concentration and is slowest (about 60 min) in the absence of tannic acid.

Table 6.2. *Preparation of a colloidal gold–protein A complex (Slot and Geuze 1981, 1985)*

1. Prepare a 1 mg/ml protein A* solution in distilled water.

2. Mix 0.25 ml of a gold sol with various amounts of protein A (e.g. 5–50 µl of a 50 µg/ml protein A solution) in small glass tubes. Add NaCl (final concentration 1%) after 1 min.

3. Change of colour from red to blue indicates that sol is not stabilized and more protein can be bound. The saturation concentration is reached when the colour does not change any more (at about 5 µg protein A/ml sol).

4. After establishing the saturation concentration for the sol (steps 1–3), add the saturation concentation of protein A to 30 ml sol, while stirring.

5. Add 0.5 ml of 0.1% bovine albumin as extra stabilizer after a few minutes.

6. Concentrate the preparation by centrifugation (e.g 5 nm gold particles for 45 min at 125 000 g_{av}, 10 nm gold particles for 45 min at 50 000 g_{av}). The pellet is composed of a loose part and a tightly packed part. Resuspend the loose part of the pellet in a smaller (1/25 of the original volume) in PBS.

7. The protein A–gold complexes can be purified by gradient centrifugation (10–30% glycerol in PBS, pH 7.2, and 0.1% BSA), e.g. 45 min at 41 000 rpm for 5 nm gold or 45 min at 20 000 rpm for 10 nm gold in SW 41 rotor (Beckman Instruments).

8. Store solutions at 277–281 K (add 0.02% sodium azide).

9. Solutions stored at 277 K may lose activity within weeks (Horrisberger and Clerc 1985). Aliquots can be stored at 203 K in 20% glycerol.

---

*Sigma or Pharmacia products can be recommended.
For recipes for making other protein–gold complexes (avidin–gold, lectin–gold etc.) see Beesley 1989.

particles of any desired size and for the coating of colloidal gold with protein A are provided in Tables 6.1 and 6.2 respectively.

Labelling can be performed on specimens prepared in several different ways, but any immunocytochemical technique requires:

(i)   specific high titre antibodies;

(ii)   that all parts of the specimen must be accessible to antibody and electron-dense marker;

(iii) that antigen–antibody binding is not affected by the fixation (i.e. that the antigen is not altered by the fixative);

(iv)   that the original fine structure is preserved as much as possible;

(v)   that multiple labelling with different electron-dense markers is feasible;

(vi)   that quantitative analysis can be undertaken.

6.2 Immunocytochemical methods: general

The 'pre-embedding' technique includes permeabilization of aldehyde fixed cells or tissue with detergents or polar solvents. The permeabilized structures are then exposed to the antibody and the electron-dense marker, refixed after the labelling procedure, and processed for conventional resin embedding and sectioning. The treatment disrupts the organization of the cell, and can lead to damage to the fine structure, and the displacement of the antigen. A further problem is that it is not possible to verify whether or not the penetration of the antibody and the marker is complete.

The 'post-embedding' technique includes either conventional epoxy resin embedding or low temperature embedding with suitable resins, and subsequent sectioning and labelling of the sections. Conventional epoxy resin embedding, however, reduces the sensitivity of labelling considerably due to the loss of antigenicity during dehydration and embedding. In addition, the ratio of specific labelling to unspecific labelling is low because the electron-dense marker tends to adhere nonspecifically to the resin. The recently developed low temperature embedding media (Lowicryl and LR resins), however, show a greatly improved sensitivity (Newman and Hobot 1987). These methods offer a distinct advantage in pathological and diagnostic contexts because the unsectioned blocks are permanent and easily stored for future reference. Since these methods are not strictly speaking low temperature methods, we would like to restrict ourselves to the description of arguably the most successful immunocytochemical method, the so-called 'Tokuyasu method' (Tokuyasu 1973), which uses thawed cryosections for the immunocytochemical localization of membrane antigens. This method:

(i)   preserves fine structure;

(ii)   has a high sensitivity;

(iii) is rapid;

(iv)   allows semithin sections to be cut from the same frozen block for labelling with fluorescent markers for light microscopy.

Wynford-Thomas *et al.* (1986) have introduced a pre-embedding/post embedding immunocytochemical method that permits the rapid correlative localization of epidermal growth factor receptors within a cell population by light microscopy, followed by a high resolution examination by electron microscopy.

We would not advocate the use of the Tokuyasu method to the exclusion of the resin-embedding procedures in all situations, although it does appear to offer advantages for many critical purposes.

6.3 Immunocytochemistry of thawed cryosections: practical aspects

6.3.1 Fixation and infusion with cryoprotectant

Tissue or cell pellets can be fixed with aldehydes using the same methods that are used for conventional resin embedding. Fixation can either be performed by immersing small tissue pieces or cell pellets in a fixative, or by perfusing whole organs with a fixative. The concentration of the fixative depends very much on the nature of the antigen. Some antigens survive fixation in for example 2% glutaraldehyde, whereas others would lose their antigenicity under such conditions and, therefore, may need to be fixed with 2% formaldehyde (Boonstra *et al.* 1985). One has to keep in mind that formaldehyde cross-linking is, to a considerable extent, reversible (Baschong *et al.* 1983), especially when using formaldehyde concentrations below 4 per cent, where formaldehyde monomers are predominant. Washing the sample after formaldehyde fixation should therefore be avoided, although the situation is less critical for higher concentrations (e.g. 8 per cent) since the proportion of polymer to monomer is higher and the cross-linking process more efficient. Cross-linking with formaldehyde is generally considered to be a slower process than with glutaraldehyde. However, since the degree of cross-linking needed for cryosections is probably less than for conventional resin embedding, the fixation time can be as short as ten minutes, depending on the composition, size, and shape (e.g. cultured cell monolayers) of the specimen. The fixed tissue blocks can be trimmed and should not be larger than 1 mm^3.

Cell or organelle suspensions are more difficult to handle. It is recommended that cell and organelle suspensions be pelleted and fixed by adding the fixative carefully onto the pellet after removal of the supernatant. After the fixation is complete, the fixative can be removed carefully. Fixed pellets are usually not very stable and tend to disintegrate when infused with sucrose. They can be stabilized by adding a substance such as gelatine to the fixed pellet at around 305 K. After infiltration with gelatine the cell/gelatine mixture is put on ice and left to solidify. Small 'tissue' blocks can be cut out and fixed once more with cold fixative.

In order to be able to section the tissue blocks it is necessary to freeze them and section them at low temperatures (Chapter 3). Since the blocks are bulk specimens it is not possible to freeze them fast enough to achieve vitrification throughout the whole block. As a result of this, ice crystals will form, thus destroying the fine structure. However, tissue blocks can be vitrified using cryoprotectants. Cryoprotectants are compounds that bind water to the extent that no water molecules are available for nucleation, the first step in ice crystal formation (Chapter 2). Fortunately chemical fixation renders cells permeable to sucrose and other short chain sugars. Infusion with high concentrations of sucrose will, therefore, protect the specimen against freezing damage.

Small fixed tissue pieces are placed in 2.1–2.3 M sucrose in phosphate buffered saline (PBS) for 30–60 min. The sucrose concentration, the degree of fixation, and the chemical composition of the specimen itself will determine the plasticity and therefore the sectioning properties of the block. The higher the sucrose concentration used, the softer the block at a given temperature. However, below 150 K, blocks containing high sucrose concentrations tend to be brittle and relatively difficult to section. Usually a higher degree of cross linking leads to better sectioning properties but reduced labelling sensitivity. A minimized fixation often results in an extensive loss of cytosolic proteins and poorer sectioning characteristics. Tokuyasu recommends, therefore, as a useful starting point with a new tissue system, that the specimen should be infused with a 1.6 M sucrose solution containing 25% w/v PVP for several hours (Tokuyasu 1986*a*). To make 100 ml of such a solution, a paste consisting of 25 g PVP and 80 ml of 2 M sucrose in 0.4 M Na_2HPO_4 is mixed in a container. The pH can be adjusted with 1 N NaOH. The container is sealed and left at room temperature overnight. The air bubbles trapped in the paste will escape leaving a clear solution. Infusion times are somewhat prolonged (several hours). In order to achieve vitrification the freezing rate has to be higher than for sucrose infused specimens. The blocks should therefore be frozen in Freon 12 or Freon 22.

6.3.2 Specimen mounting and freezing

The trimmed tissue pieces are mounted on to the specimen support (silver, aluminium, or copper stubs/pins) that fits the specimen holder of the microtome. The specimen should be mounted so that the smallest surface available faces the knife in the microtome. Excess sucrose is removed with filter paper but the specimen should not be allowed to dry out. The specimen support and the attached specimen are plunged into liquid nitrogen for freezing. If the sucrose (in the 2.1–2.3 M concentration range) has penetrated the specimen completely, plunge freezing in liquid nitrogen is sufficient to vitrify the specimen (Griffiths *et al.* 1984), and more sophisticated, faster freezing methods would not improve the freezing quality.

6.3.3 Sectioning and section retrieval

Sectioning will obviously have to take place in a cryoultramicrotome. Sections for light microscopical immunofluorescence studies (section thickness 0.2–1.0 μm) can be cut at temperatures between 203 K and 243 K, and sections for electron microscopical studies between 163 K and 193 K (details about sectioning devices, the sectioning process, and the knives are to be found in Chapter 3). Sections have to be removed from the knife edge and assembled in small groups (3–5 sections) for picking up. This can be done using an eyelash probe. Sometimes static electricity can cause the sections to stick to each other or jump to the nearest surface. Anti-static devices can help to reduce the problem. To pick up a group of sections a drop of 2.3 M sucrose in PBS is taken up with a copper or platinum wire loop (1–2 mm diameter) and then moved into the cryochamber and brought very close to the sections. The velocity at which the loop is moved towards the knife (i.e. through the chamber atmosphere) and the size of the drop determine the final temperature of the drop when it comes into contact with the sections. The drop should touch the sections just before it freezes completely. If the drop is too fluid the sucrose might spread over the knife, and if it is frozen the compressed setions will not stretch properly even upon subsequent melting. When sectioning for light microscopy at around 233 K, the loop with the drop should be left in the cryochamber for a short period of time until the appearance of a white freezing spot on the drop surface indicates that the drop is cold enough. When sectioning for electron microscopical studies (e.g. at 173 K) one should avoid touching the surface of the knife with the drop, since it would cool the drop even more rapidly and cause water to condense on to this part of the knife. It is, therefore, sufficient to approach only as close as 1 mm from the sections, at which point they will be attracted by the drop.

The drop with its suspended sections is then taken out of the cryochamber and deposited on a glass slide, or on a grid covered with a glow-discharged hydrophilic support film. Usually more than one grid will be used for the subsequent labelling steps. During the collection of additional grids, the grids with sections attached to them can be stored in a number of different ways, for example:

1. on drops of PBS (phosphate buffered saline) containing 5% new-born calf serum or fetal calf serum on ice;
2. on 2% gelatine in PBS within small petri dishes placed on ice;
3. on 0.3% agarose mixed with 2% gelatine.

The grids will have to float on the substrate with the sections facing the liquid phase. Extreme care has to be taken to ensure that at no time during the procedure are the sections allowed to dry out. The grids should not be immersed in the solutions.

6.3.4 The labelling or immuno-staining procedure

The entire labelling sequence can be performed at room temperature by transferring the grids section-side down through a series of droplets arranged on a fresh sheet of parafilm. The back side of the grid should stay dry. The transfer can be done with a 3 mm wire loop or a pair of curved tweezers. A schedule for the labelling sequence with an antibody and protein A–gold for electron microscopy is summarized in Table 6.3, and a schedule for the immunofluorescence labelling procedure for light microscopy in Table 6.4. Gold labelling patterns are usually too faint to be observed in the light microscope. With silver enhancement (Danscher and Norgaard 1983; Holgate *et al.* 1983) each individual gold particle can be increased in size and visualized in the light microscope (Slot *et al.* 1986). The method has some

Table 6.3. *Labelling procedure for light microscopy**

1.	Glass slide preparation	Mark an area on the glass slide with a diamond scorer. Coat the degreased slide with either gelatine (1% gelatine + 0.1% potassium chromosulphate) or polylysine (leave slide for at least 30 min in a 0.01% solution in distilled water) and rinse with double distilled water 2–3 times.
2.	Sectioning	Section at 203 K slightly thicker than for electron microscopy (around 200 nm).
3.	Section collection	Pick up sections with a loop containing a drop of a 2.3 M sucrose solution. Touch marked area on the slide with the loop. The sections will stick to the slide.
4.	Fixing sections to slide	Immerse slide in a Coplin jar containing 3% formaldehyde in PBS. Wash once in PBS.
5	Quenching of aldehyde groups	Immerse slide in 50 mM NH_4Cl/PBS for 10 min. Wash twice in PBS for 3 min.
6.	Antibody-binding	Rinse in 0.2% gelatine/PBS. Remove excess fluid, but take care not to dry the sections. Add antibody, dilute in 0.2% gelatine/PBS to the sections and incubate for 15–30 min.
7.	Washing	Wash twice in a Coplin jar containing PBS for a total of 10 min
8.	Second antibody	Repeat (6) with fluoresceinated antibody for 15–30 min (or protein A–gold).† (Add protein A–gold for 30 min.)
9.	Washing	Wash twice in PBS for a total of 10 min. Wash twice in distilled water for a total of 6 min.
10.	Mounting	Mount the slides with either Moviol or Elvanol.

*Although not strictly within the context of this electron microscopical text, it is necessary to consider light microscopic labelling procedures because (a) of the intrinsic value and perspective that the information provides that complements the EM observations, (b) they provide a relatively quick and directly pertinent means of evaluating antibodies.
†If protein A–gold is used, a subsequent silver enhancement procedure has to be employed (see Table 6.5).

important advantages over the fluorescence technique, namely that the preparation does not fade, no fluorescence microscope is required, and there is no problem with autofluorescence. A silver enhancement procedure is summarized in Table 6.5.

Table 6.4. *Labelling procedure for electron microscopy*

1.	Grid collection	Collect grids on 10% fetal calf serum (FCS)/PBS on ice; wash in PBS once before proceeding.
2.	Quenching	If the cells have been glutaraldehyde fixed, free aldehyde groups can be quenched in 0.02 M glycine in 5–10% FCS/PBS for 10 min; wash twice in PBS for a total of 5 min.
3.	Antibody-binding	Spin antibody solution (1 min at 13 000 g, dilute in 5–10% FCS/PBS; incubate sections for 15–60 min.
4.	Washing	Wash six times in PBS for a total of 15 min.
5.	Protein A-gold	Incubate grids in protein A–gold for 20–30 min. Dilute protein A–gold in 5–10% FCS/PBS. The concentration is critical. Too high a concentration gives non-specific binding.
6.	Washing	Wash six times in PBS for a total of 25 min. Wash four times in distilled water for a total of 5 min.
7.	Staining	Stain with 2% uranyl acetate/oxalate at pH 7–7.5. (Mix equal parts of 0.3 M oxalic acid and 4% aqueous uranyl acetate; adjust pH with 5% NH_4OH.)
8.	Washing	Wash twice in distilled water for a total of 3 min.
9.	Embedding	Embed three times with 2% methyl cellulose solution (25 cp) containing 0.1–0.4% uranyl acetate for 10–20 min.
10.	Drying	Pick the grid up with a 3 mm loop and remove the excess fluid with filter paper. Dry the grid suspended in the loop in a chamber containing silica gel for 5 min. The thickness of the methyl cellulose film determines the contrast and the extent of drying artefacts.

Table 6.5. *Silver enhancement of protein A–gold labelled sections for light microscope examination (Slot et al. 1986)*

1.	Immunostaining	Sectioning, section collection, and immunostaining as described in Table 6.3, 1–9.
2.	Developing	Place the slide in developer containing 0.85 g hydroquinone and 0.11 g silver lactate per 100 ml sodium citrate buffer at pH 4, for 3–4 min.
3.	Washing	Rinse slide with distilled water.
4.	Fixation	Fix for one minute in diluted (1:10) photographic fixer (e.g. Ilfospeed 2000, Ilford, UK).
5.	Washing	Rinse slide in distilled water.
6.	Counterstaining	Sections can be counterstained with hematoxylin.
7.	Mounting	Mount in either Moviol or Entellan (Merck).

Easy to use silver enhancement kits are commercially available (e.g. IntenSE kit, Janssen).

6.3.5 Contrasting and drying of the labelled sections

In order to maintain the fine structure, and to obtain reasonable contrast, this final step is crucial (Griffiths *et al.* 1984). The sections are postembedded in a relatively high concentration (2 per cent) of a low-viscosity methyl cellulose (25 centipoise) containing uranyl acetate. With uranyl acetate concentrations above 0.3 per cent, increased negative stain effect is obtained. The final thickness of the methyl cellulose is critical with respect to both contrast and preservation of the fine structure. A thick film will improve the preservation of fine structure but reduce the contrast. Very thin films will give high contrast but lead to significant air-drying artefacts.

A recent modification (Tokuyasu 1986*b*) of section embedding and staining is based on the use of polyvinylalcohol (PVA) instead of methyl cellulose. Advantages of PVA are: (i) it has a low molecular weight (10 000) and can therefore penetrate better into the sections, (ii) it is less viscous than methyl cellulose and concentrations as high as 4 per cent can be used, and (iii) it can be applied in combination with lead staining.

6.3.6 Applications

Thawed cryosections have been used successfully to localize albumin in rat liver parenchymal cells (Brands *et al.* 1983), epidermal growth factor receptor in cultured A432 cells (Boonstra *et al.* 1985), serum albumin in the rat testis (Christensen *et al.* 1985), and pupal cuticle proteins in *Drosophila melanogaster* (Wolfgang *et al.* 1986). They have also been used to study the fate of the surface coat of the parasite *Theileria parva* after the entry of the sporozoites into bovine lymphocytes (Webster *et al.* 1985). The method is also contributing considerably to our understanding of intracellular membrane traffic and sorting events (Green *et al.* 1987; Griffiths *et al.* 1988; Griffiths *et al.* 1989, 1990), and has been used to correlate immunogold labelling intensities to absolute numbers of membrane antigens (Griffiths and Hoppeler 1986; Posthuma *et al.* 1987).

7 Other low temperature methods

7.1 Freeze-drying

7.1.1 General principles

For the examination of biological material in the electron microscope at ambient temperatures, water must either be removed from the specimen, or immobilized. Dehydration can be achieved in several ways, for example air-drying, critical point drying, or freeze-drying (Robards and Sleytr 1985). Freeze-drying, a well established and versatile method, is defined as the removal of water vapour from ice. The sublimation of water molecules from the solid phase (ice) depends on the temperature of the ice and can only occur when the partial pressure of the water vapour in the frozen surface exceeds that of the atmosphere immediately adjacent to it. The reverse process, the condensation of water molecules on to the ice, depends on the pressure and temperature of the vapour phase, It is very difficult to calculate theoretical sublimation rates, although Newbury et al. (1986) provide a useful expression:

$$Q_n = 0.25(P_T - P_A) T^{0.5} \tag{7.1}$$

where Q_n is the amount of water lost from the specimen (g cm^{-2} s^{-1}, P_A is the ambient vapour pressure (Pascals) of water, P_T is the saturated vapour pressure (Pascals) of water at the specimen temperature, T(K). (N.B. when the vacuum pressure is sufficiently low, or when the specimen is surrounded by a sufficiently dry atmosphere, Q_n will be determined by T. For example, at 273 K the maximum evaporation rate from pure ice is 8.3×10^{-2} g cm^{-2} s^{-1}, whilst at 173 K the rate is 5×10^{-7} g cm^{-2} s^{-1}, i.e. 100 000 times slower.) Umrath (1983) used this expression to estimate the decrease of a pure ice layer in relation to the sublimation time, and Table 7.1 provides an edited version of his results.

In biological tissue specimens the situation is more complex. Sublimation of water leads to a layer of dry biological material which inhibits further sublimation. For this reason the drying times for a tissue specimen of a given thickness are 20–100 times longer than for solid pure water of similar thickness (Umrath 1983). Furthermore, biological specimens vary in size, structure, and composition, and it is therefore difficult to establish a general freeze-drying regime.

Table 7.1. *Comparison of freeze-drying times (data taken from Umrath 1983)*

Temperature (K)	Vapour pressure (10^{-3} Pa)	Drying time (s/μm)	Drying time per μm			
			d	h	min	s
213	1080	0.676				0.68
203	259	2.75				2.75
193	53.6	13				13
192	45.3	15.3				15
191	38.3	18.1				18
190	32.3	21.4				21
189	27.2	25.3				25
188	22.9	30.0				30
187	19.2	35.7				36
186	16.0	42.5				42
185	13.5	50.8				51
184	11.2	60.7			1	1
183	9.32	78.2			1	18
173	1.32	499			8	19
163	0.148	4340		1	12	20
153	0.0124	50200		13	56	40
143	0.000739	817000	9	10	56	

7.1.2 Freeze-drying of biological specimens and resin embedding

Biological specimens can be frozen-dried and subsequently embedded in resin. Even though this method undoubtedly introduces its own artefacts, it has been developed as an alternative to the conventional wet-chemical preparation methods, since it offers a more valid preservation of the original specimen composition and structure (Roos and Barnard 1986; Roos *et al.* 1990). In the study of the cytology of heavy metals, this technique permits a high throughput of specimens containing cadmium (Nott and Langston 1989). Resin embedded specimens can be stored indefinitely.

Biological specimens are frozen rapidly in order to avoid ice crystal growth. The freezing rate will determine the quality of the freezing. Low freezing rates will result in hexagonal ice in the specimen and the destruction of fine structural details (Chapter 2). With very small specimens and high freezing rates, vitrification can be achieved. However, vitrification can only be maintained at temperatures below the devitrification or recrystallization point. This recrystallization temperature is 135 K for pure water and probably around 193 K for soft tissues. Freeze-drying should ideally be performed at temperatures lower than the recrystallization temperature in order to avoid both the formation and growth of ice crystals during the drying process. This, however, would result in impracticably long freeze-drying schedules (Table 7.1). In order to get a reasonable sublimation rate it is necessary to choose a drying temperature above the recrystallization temperature of the specimen. Experiments using tissue models (Stumpf and

Roth 1967) show clearly that the sublimation rate decreases exponentially with the thickness of the already dry surface layer, indicating that freeze-drying is most effectively performed on specimens which are as small as possible.

7.1.3 Freeze-drying apparatus for small bulk specimens

Best freeze-drying results are obtained when the partial pressures of the gases around the specimen are kept much lower (preferably by a factor of about 1000) than the saturation vapour pressure for ice at the specimen temperature. Three different approaches are available to remove the released water vapour and other gases from the system:

(1) *condensation* of the vapour on to a surface (condenser, cold finger) kept at a lower temperature than the specimen;

(2) *pumping* the vapour using rotary, oil diffusion or turbo molecular pumps;

(3) *adsorption* of the vapour on to a chemical (phosphorus pentoxide, calcium chloride, copper sulphate, or a molecular sieve).

Most 'home made' and commercially-available freeze-drying units combine at least two of the strategies. Systems that combine pumping and condensers or chemical desiccants should have the condenser or the chemical closer to the specimen than the mean free path of the residual gases at a given pressure to ensure maximum removal of molecules from the gas phase.

The mean free path is given by:

$$\lambda = 1/\sqrt{(2\pi n d^2)} \tag{7.2}$$

where λ is the mean free path (the average distance a molecule would have to travel before striking another molecule), d the diameter of the molecule, and n the number of molecules per cm^3.

If molecular sieves are used at low temperatures they perform the functions of a vapour trap and vacuum pump, and they can be reactivated simply by heating. Edelmann (1978) describes a simple system based on the use of molecular sieve that operates without additional vacuum pumps.

All vacuum-pumping freeze-drying systems should have the following features (a) a vacuum gauge, (b) temperature control of the specimen holder, and (c) a thermocouple close to the specimen to monitor the actual specimen temperature. Two freeze-drying units are schematically presented in Fig. 7.1. A general freeze-drying and resin-embedding regime is summarized in Table 7.2. Most conventional resins are suitable provided their viscosity is low enough to ensure complete infiltration of the specimen. Some resins allow infiltration and polymerization at low temperatures (Section 7.3). The morphological preservation of tissues by the freeze-drying/resin embedding

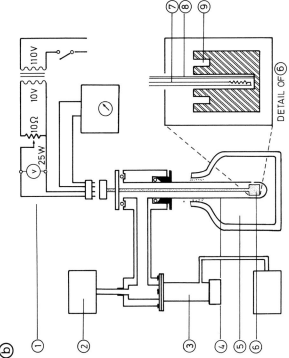

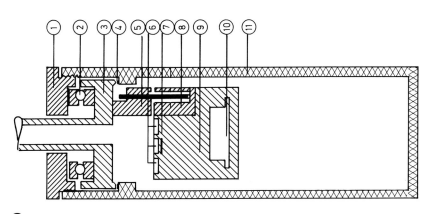

Fig. 7.1. Two relatively simple types of freeze-drying apparatus. (a) Container is immersed in liquid nitrogen. The vacuum is achieved by cryosorbtion (molecular sieve) and the cold walls act as a condenser (redrawn from Edelmann 1978); 1 = screw cover, 2 = ball bearing, 3 = cover (with tube), 4 = washer, 5 = plug, 6 = specimen holes, 7 = thermocouple, 8 = socket, 9 = specimen support, 10 = heating element, 11 = vessel. (b) The freeze-drying unit is immersed in liquid nitrogen, but the vacuum is provided by a rotary pump (redrawn from Stirling and Kinter 1967); 1 = power supply, 2 = pressure gauge, 3 = rotary pump, 4 = vessel, 5 = liquid nitrogen containing dewar, 6 = specimen support, 7 = heating element, 8 = thermocouple, 9 = specimen holes.

method can be very good (Fig. 7.2), but the technique is subject to a number of artefacts:

1. The formation of ice crystals during freezing and secondary ice crystal growth due to freeze-drying temperatures necessarily higher than the critical recrystallization temperature may cause severe morphological changes (Chapter 2).

2. Structures subjected to freeze-drying may shrink and collapse.

Table 7.2. *A typical freeze-drying schedule for quench-frozen soft biological tissue (Roos and Barnard 1985)*

Accumulated time (h)	Drying time (h)	Temperature (K)
1	1	138
3	2	138–203
33	30	203
57	24	203–233
63	6	233–297
80	17	297
Formaldehyde-vapour fixation	(3)	333
Osmium-vapour fixation	(3–18)	297
Plastic-infiltration	(2–4)	297

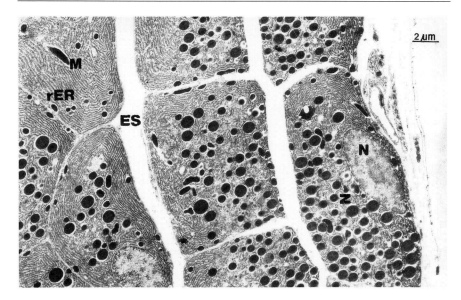

Fig. 7.2. Transmission electron micrograph of an ultrathin section of quench-frozen, freeze-dried, osmium-vapour-fixed, plastic-embedded rat exocrine pancreas. Section was post-stained with uranyl acetate/lead citrate to illustrate the quality of structural preservation. Such a specimen is of little use for the X-ray microanalysis of solutes because of leaching losses during staining. M = mitochondria, N = nucleus, rER = rough endoplasmic reticulum, Z = zymogen granules, ES = extracellular space.

3. Frozen-dried specimens are usually extremely hygroscopic. Rehydration can be avoided by keeping the specimen at temperatures slightly higher than the environment.

4. Infiltration of the dry specimen with a resin may cause redistribution of elements (Chapter 4).

7.2 Freeze-substitution

7.2.1 General principles

Freeze-substitution is a procedure whereby the ice in a quench-frozen specimen is slowly replaced at low temperatures by anhydrous organic solvent molecules, followed by infiltration with a resin. The method has been used in electron microscopy since 1957 (Fernández-Morán 1957) and has been modified and improved ever since. In common with all cryopreparative procedures, the final result is very much dependent on the quality of the freezing method. Assuming that the quality of freezing is good, the secondary objective is to substitute the ice at temperatures below the recrystallization point of the specimen. The substitution medium should therefore be liquid below that temperature (193 K), and should ideally have the capacity to dissolve additives such as OsO_4 for fixation and staining of specimen structures. Some frequently used freeze-substitution liquids and their properties are listed in Table 7.3.

Table 7.3. *Some commonly used freeze-substitution liquids (Robards and Sleytr 1985)*

Solvent	Melting point (K)
Ethanol	156
Diethyl ether	157
Propylene oxide	161
Acetone	178
Methanol	179
Acrolein	186

The quality of the freeze-substitution, however, is dependent not only on the choice of the substitution medium, but also on the conditions during the substitution process. For some applications (e.g. X-ray microanalysis) water has to be removed from the substitution liquid before specimen immersion and constantly during the substitution phase (Marshall 1980). This is achieved by adding molecular sieves to the system. Another parameter affecting the quality of the substitution process is the time-course of the warming up after the substitution has taken place. Using model systems labelled with tritiated water, and methanol, diethylether, and acetone as substitution liquids, Humbel *et al.* (1983) showed that methanol is the fastest substitution liquid, and diethylether the slowest.

Whatever substitution medium is chosen, for all critical purposes where soluble constituents are to be localized, it is recommended that molecular sieves are used to capture the water released during substitution and embedding.

7.2.2 Freeze-substitution equipment

The required hardware is easily made using basic workshop facilities and/or some imagination. It is, however, fairly important that the device permits the temperature of the specimen and the substitution medium to be measured and controlled in the range 173–294 K. In addition, defined temperatures should be stable over a set period of time (usually several hours, or a few days), and the whole arrangement should be sealed off from the ambient atmosphere, to ensure that the water content of the substitution medium is maintained as low as possible.

A simple freeze-substitution set-up consists of a metal container filled with substitution liquid cooled by liquid nitrogen. The frozen specimen is placed on the frozen surface of the substitution liquid. The metal container is insulated, with the insulator determining the rate of warming. It is not possible to keep a defined temperature for a period of time with this system. More sophisticated methods including temperature control (in the range 77–273 K) and presetting of times (up to 100 days) are described by Müller (1981) and Sitte (1984).

Freeze-substitution devices are also commercially available. Reichert–Jung (Cambridge Instruments) sell the 'CS auto', a device manufactured according to principles defined by Sitte and Edelmann (Sitte 1984). RMC market freeze-substitution systems called MS 6000 and MS 6200, and Balzer's market a system designated FSU 010.

7.2.3 Freeze-substitution regimes for thin sectioning

For the preservation of morphology by freeze-substitution, the major problem is to fix cellular structures after the ice has been replaced to prevent excessive extraction of cellular material. However, fixatives dissolved in the substitution medium, such as aldehydes and OsO_4, are usually not active at very low temperatures. Substitution schedules therefore include a slow step-wise temperature increase after the actual substitution has taken place. A general freeze-substitution regime is presented in Table 7.4. If thin sections of freeze-substituted material are to be used for the localization of electro-lytes, a more complex schedule has to be employed to minimize the risk of redistribution and loss of elements (Harvey 1982; Marshall 1980). Morpho-

Table 7.4. *A general freeze-substitution regime for morphological studies*

1. Rapid freezing	Of preferably small tissue specimens
2. Transfer	Of the quench-frozen specimen on to frozen substitution liquid at 77 K
3. Substitution	While shaking at 183 K for 8–12 h depending on specimen size and volume
4. Warming	To 213 K for 8 h To 243 K for 8 h
5. Washing	Twice in substitution liquid kept at 243 K
6. Embedding	Using conventional epoxy resins (such as Spurr's or Epon) or 'low temperature resins' such as the Lowicryls and the LR White/Gold resins

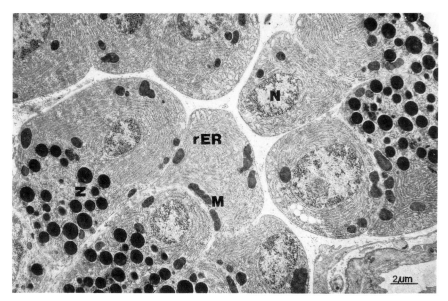

Fig. 7.3. Transmission electron micrograph of an ultrathin section of rat exocrine pancreas freeze-substituted in acetone/osmium mixture. The section was post-stained in uranyl acetate/lead citrate, and the micrograph taken about 50 μm deep into the tissue (i.e. measured from the freezing-front). (Compare the morphology with Fig. 7.2). M = mitochondria, N = nucleus, rER = rough endoplasmic reticulum, Z = zymogen granule.

logical preservation with this technique can be impressive, particularly if a fixative is incorporated into the substitution medium (Fig. 7.3).

The utility and validity of this technique for X-ray microanalytical purposes are not, however, unequivocally proven (Roos and Barnard 1986), so it would not be reasonable to recommend it in preference to cryomicrotomy for quantitative purposes. Freeze-substitution is frequently used as an adjunct to other more definitive cryoprocedures for morphologically assessing the quality of freezing and the extent of freezing damage. It is also frequently used for botanical studies (Harvey 1982), where the plant material often does not permit cryosectioning because of large water and air spaces. Recent studies show, however, that certain plant tissues can be cryo-sectioned (Parsons *et al.* 1984), so we can anticipate more concerted efforts in this direction in the future.

7.3 Low temperature embedding

The development of resins (methacrylate and acrylic formulations) with low enough viscosity to allow for infiltration and polymerization at temperatures below 273 K (Carlemalm *et al.* 1982; Newman and Hobot 1987) have opened a new field of applications for the freeze-substitution and freeze-drying methods. The resins used mostly are the methacrylates Lowicryl K4M and K11M, polar resins usable at 237 K and down to 190 K respectively, and Lowicryl HM20 and HM23, nonpolar resins applicable at temperatures as low as 190 K (Carlemalm *et al.* 1985*b*). Both types of resin are relatively stable to electron beams, and can be polymerized either by heat at 333 K or by UV light (360 nm) at low temperatures (e.g. in a conventional refrigerator). If UV light is used, vapour fixation with OsO_4 must be avoided, since osmium blocks the penetration of the UV rays. Providing that the penetration of the resin into the specimen and the polymerization of the resin is complete, the plastic can be sectioned with either glass or diamond knives. In the case of the polar K4M and K11M resins, contact between the trough liquid and the block face must be avoided.

The resins have mainly been used in conjunction with the progressive lowering of temperatures, thus replacing conventional embedding techniques (for a schedule see Table 7.5), and to perform immunocytochemical labelling on sectioned material (Roth 1983; Causton 1984). It is claimed that these embedding procedures reduce the extraction of biological material and the denaturation of antigenic sites. However, for some antigens the yield of labelling sites was not as good as expected after embedding in K4M (Roth *et*

Table 7.5. *Embedding* * *with Lowicryl K4M resin*

Step	Medium	Temperature (K)	Time
1. Dehydration	30% ethanol	273	30 min
	50% ethanol	258	60 min
	70% ethanol	248	60 min
	100% ethanol	238	120 min
2. Infiltration	1:1 100% ethanol and K4M	238	60 min
	1:2 100% ethanol and K4M	238	60 min
	100% K4M	238	60 min
	100% K4M	238	overnight
3. Polymerization (UV lamp)	100% K4M	238	24 h
		295	72 h

*Follow the manufacturer's guidelines, or those provided by Glauert and Young (1989).

al. 1981) and the results were often not reproducible. This could be the result of a rise in temperature during polymerization, which is an exothermic reaction (Ashford *et al.* 1986), although the temperature rise can be controlled or eliminated completely if certain precautions are taken (Glauert and Young, 1989). The more important precautions are (a) the use of diffuse UV radiation and (b) the provision of heat 'sinks' (i.e. containers for flat embedding are placed in recessed aluminium blocks, and gelatin capsules are held in an ethanol bath). In a few studies, freeze-substituted material (Ebersold *et al.* 1981; Humbel and Müller 1986) and freeze-dried specimens (Wroblewski *et al.* 1985; Edelmann 1986) have been infiltrated at low temperatures.

Careful control of the polymerization of the acrylic, LR White (Newman *et al.* 1982, 1983) confers versatility and certain attractive features (Newman and Hobot 1987) to the resin in the contexts of sub-ambient embedding temperatures and the immunolocalization of antigens:

1. They are electron-beam stable.
2. They are hydrophillic, so background staining with equally hydrophilic immunoreagents is significantly reduced.
3. LR White is compatible with up to 12 per cent tissue water, so partial dehydration is permissible, thus improving antibody–antigen reaction.
4. LR White is less lipophilic than epoxides and is therefore less likely to disrupt cell ultrastructure when osmication is omitted.
5. Correlative immunolabelling for light and electron microscopy can easily be achieved on the same specimen—a considerable advantage, especially in the context of the examination of valuable, perhaps unique, pathological specimens.

Another new formulation, the LR Gold resin has been introduced recently. It allows infiltration of the unfixed specimen at 248 K. Polymerization can take place at the same temperature using visible light and is complete after 24 hours.

7.4 Other cryotechniques for X-ray microanalysis

We have already discussed in Chapters 3, 4, and 5 the techniques for preparing thin frozen-hydrated and frozen-dried cryosections suitable for X-ray microanalysis. Thin frozen-dried cryosections must be considered the specimens of choice for most biological microprobe purposes, until our ability to examine hydrated specimens matches our ability to produce them. However, it is only fair to add that resin-embedded freeze-dried and freeze-substituted thin sections are advocated by certain workers (see reviews by Wroblewski and Wroblewski 1986; Wroblewski *et al.* 1988), although the difficulty of validating the techniques and, thus, of recommending their usage must be emphasized (Roos and Barnard 1985, 1986).

For low-resolution microprobe analyses, conducted on a whole-cell or nucleus/cytoplasm level, thick (~ 16 μm) or semi-thick (2–6 μm) freeze-dried sections cut on a histological cryostat (temperature range $- 30$ °C to $- 35$ °C) can be realistic options (Cameron *et al.* 1979; Wroblewski *et al.* 1988), providing the objectives of the investigations are of modest ambitions. Such preparations have been used because (a) a large population of cells can be analytically sampled very quickly and with relative ease, and (b) in surgical and pathological contexts, where the opportunity for optimal freezing is often precluded, but where the inevitable gross ice-damage is not prohibitive.

8 Safety considerations and concluding remarks

We have repeated several points, occasionally to the extent of labouring them, in this short book. Where this has happened it has usually been deliberate, and for two (hopefully excusable) reasons. First, because cryo-electron microscopy is a multidisciplinary activity, it seems to us that, in order to encourage a healthy and productive cross-fertilization of ideas and procedures, fundamentally important aspects (such as freezing and radiation damage) should be presented in a number of different contexts, and not packaged for the consumption of a single cohort of practiners. Second, some of these fundamental points contain the keys to eventual success, and for that reason they bear repetition.

Before closing this handbook it behoves us to draw attention to the very important issues of *safety in the cryogenics laboratory*. Given space constraints, we have only compiled a list of the more important aspects here. The reader is strongly urged to consult the two publications (Robards and Sleytr 1985; Ryan and Liddicoat 1987) upon which this list is based.

1. Working at low temperatures is potentially dangerous! The danger is compounded in the cryo-e.m. laboratory because hazards are presented not only by the cryogen and cold metal surfaces. Vacuum equipment, high pressure vessels, high voltages, inflammable organic solvents, toxic chemicals and ionizing radiation are also hazardous.

2. Many common (and otherwise excellent) cryogens are explosive! Both liquid propane and ethane are potentially very dangerous.

3. When handling liquid cryogens it is essential to protect the eyes (in particular) and the face with either goggles or a full face shield. Sensible clothing and footwear should be worn—these should be tight fitting and dry, the object being to prevent spillages collecting in pockets, etc. Gloves should be loose-fitting for easy removal.

4. Personnel should be aware of first-aid procedures in the event of a cryogenic accident. Minor burns should preferably be treated by immersion in lukewarm (or, in an emergency, cold) water and then by application of a cold compress. More serious accidents require immediate medical attention.

5. Flammable gases must be stored outdoors.

6. Liquefaction should be performed in a fume cupboard.

7. Gas condensation (by liquid N_2 cooling) should be slow and the end of the gas delivery tube should not be immersed under the newly condensed liquid in the receptacle.

8. Propane gas should be ducted from the storage cylinder through coils of narrow copper pipe immersed in liquid N_2 so that the propane is then delivered slowly from the pipe at a temperature below its flash point of 169 K (143 K for ethane). The coolant should always be maintained under an inert nitrogen atmosphere derived from liquid N_2.

9. No smoking or naked lights in the vicinity of cryogenic working areas. Isolate possible ignition sources (e.g. motors) from the cryogen wherever possible.

10. Condense the cryogens in mechanically stable containers to minimize the risk of accidental spillages.

11. Dispose of the coolants carefully. The simplest method is to evaporate them in the fume-cupboard. Alternatively, pour on to a hard surface outdoors and away from ignition sources.

12. Maintain good ventilation in all cryogenic working areas.

13. Display clear hazard notices near the working area.

We must close this discourse with the simple wish that the reader has formed the firm impression that cryo-electron microscopy provides the means of studying the structure and composition of cells captured during a momentary phase in an ongoing sequence of multifarious dynamic processes, such as macromolecular conformation changes, membrane trafficking and fusions, transcellular transport of materials, ion shifts, etc. (Plattner and Knoll 1987). Cryo-electron microscopy thus superimposes a time resolution on to the spatial resolution capability historically associated with biological electron microscopy. Little wonder, therefore that P. Echlin in an editorial introducing papers presented to the Third International Low Temperature Biological Microscopy and Analysis meeting in Cambridge, April 1985, was moved to remark: '. . . the arguments for adopting cryotechniques and low temperature microscopy and analysis are secure and proven. While it should come as no surprise that so many people are now using one or more of these methods for studying biological systems, it is astonishing that any one should continue operating an electron beam instrument at ambient temperatures' (Echlin 1985).

Whilst not wishing to denigrate the contributions made by conventional procedures towards a better understanding of cell structure and function, the burgeoning application of cryotechniques is bound to continue in the future to furnish a wealth of important new information and insights.

Welcome to the field of electron microscopy where, to use Coleridge's words out of context, 'the frost performs its secret ministry'.

References

Adrian, M., Dubochet, J., Lepault, J., and McDowall, A. W. (1984). Cryo-electron microscopy of viruses. *Nature*, **308**, 32–6.

Agar, A. W., Alderson, R. H., and Chescoe, D. (1974). In *Principles and practice of electron microscope operation: practical methods in electron microscopy* (ed. A. M. Glauert). North Holland Publishing Company, Amsterdam, New York, Oxford.

Åkerström, B. and Björck, L. (1986). A physicochemical study of protein G, a molecule with unique immunoglobulin G-binding properties. *J. Biol. Chem.*, **261**, 10240–7.

Appleton, T. C. (1974). A cryostat approach to ultrathin 'dry' frozen sections for electron microscopy: a morphological and X-ray analytical study. *J. Microsc.*, **100**, 49–74.

Ardenne, M. V. (1939). Die Keilschnittmethode, ein Weg zur Herstellung von Mikrotomschnitten mit weniger als 10^{-3} mm Stärke für elektronenmikroskopische Zwecke. *Z. Wiss. Mikrosk.*, **56**, 2–23.

Ashford, A. E., Allaway, W. G., Gubler, F., Lennon, A., and Sleegers, J. (1986). Temperature control in Lowicryl K4M and glycol methacrylate during polymerization: is there a low-temperature embedding method? *J. Microsc.*, **144**, 107–26.

Bachmann, L. and Schmitt, W. W. (1971). Improved cryofixation applicable to freeze-etching. *Proc. Natl. Acad. Sci. USA*, **68**, 2149–52.

Bahr, G. F., Johnson, F. B., and Zeitler, E. (1965). The elementary composition of organic objects after electron irradiation. *Lab. Invest.*, **14**, 1115–33.

Baker, J. R. J. and Appleton, T. C. (1976). A technique for electron microscopy autoradiography (and X-ray microanalysis) of diffusible substances using freeze-dried fresh frozen sections. *J. Microsc.*, **108**, 307–15.

Bald, W. B. (1987). *Quantitative cryofixation*. Adam Hilger, Bristol and Philadelphia.

Barnard, T. (1982). Thin frozen-dried cryosections and biological X-ray microanalysis. *J. Microsc.*, **126**, 317–32.

Barnard, T., Gupta, B. L., and Hall, T. A. (1984). Effects of the high molecular weight cryoprotectant dextran on fluid secretion by *Calliphora* salivary glands. *Cryobiology*, **21**, 559–69.

Baschong, W., Baschong-Prescianotto, C., and Kellenberger, E. (1983). Reversible fixation for the study of morphology and macromolecular composition of fragile biological structures. *Eur. J. Cell Biol.*, **32**, 1–9.

Beesley, J. E. (1989). *Colloidal gold: a new perspective for cytochemical marking*. Royal Microscopical Society/Oxford University Press, Oxford.

Bendayan, M. (1982). Double immunocytochemical labelling applying the protein A–gold technique. *J. Histochem. Cytochem.*, **30**, 81–5.

Bendayan, M. (1987). Introduction of the protein G–gold complex for high resolution immunocytochemistry. *J. Electron Microsc. Tech.*, **6**, 7–13.

Bernhard, W. (1965). Ultramicrotomie à basse température. *Année Biol.*, **4**, 5–19.

Boonstra, J., van Maurik, P., Defize, L. H. K., de Laat, S. W., Leunissen, J. L. M., and

Verkley, A. J. (1985). Visualization of epidermal growth factor receptor in cryo-sections of cultured A431 cells by immuno-gold labelling. *Eur. J. Cell Biol.*, **36**, 209–16.

Brands, R., Slot, J. W., and Geuze, H. J. (1983). Albumin localization in rat liver parenchymal cells. *Eur. J. Cell Biol.*, **32**, 99–107.

Bruggeller, P. and Meyer, E. (1980). Complete vitrification in pure liquid water and dilute aqueous solutions. *Nature*, **288**, 569–71.

Butler, E. P. and Hale K. F. (1981). In *Dynamic experiments in the electron microscope: practical methods in electron microscopy*, Vol. 9 (ed. A. M. Glauert). Elsevier/North-Holland, Amsterdam.

Cameron, I. L., Smith, N. K. R., and Pool, T. B. (1979). Element concentration changes in mitotically active and postmitotic enterocytes. An X-ray microanalysis study. *J. Cell Biol.*, **80**, 444–50.

Cantino, M. E., Wilkinson, L. E., Goddard, M. K., and Johnson, D. E. (1986). Beam induced mass loss in high resolution biological microanalysis. *J. Microsc.*, **144**, 317–27.

Carlemalm, E., Garavito, R. M., and Villiger, W. (1982). Resin development for electron microscopy and analysis of embedding at low temperature. *J. Microsc.*, **126**, 123–43.

Carlemalm, E., Collnix, C., and Kellenberger, E. (1985*a*). Contrast formation in electron microscopy of biological material. *Adv. Electr.*, **63**, 269–334.

Carlemalm, E., Villiger, W., Hobot, J. A., Acetarin, J. D., and Kellenberger, E. (1985*b*). Low temperature embedding with Lowicryl resins: two new formulations and some applications. *J. Mol. Biol.*, **140**, 55–63.

Causton, B. E. (1984). The choice of resins for electron immunocytochemistry. In *Immunolabelling for electron microscopy* (ed. J. M. Polak and I. M. Varndell), pp. 29–36. Elsevier Science Publishers, Amsterdam.

Chang, J.-J., McDowall, A. W., Lepault, J., Freeman, R., and Walter, C. A. (1983). Freezing, sectioning and observation artefacts of frozen-hydrated sections for electron microscopy. *J. Microsc.*, **132**, 109–23.

Christensen, A. K. (1971). Frozen thin sections of fresh tissue for electron micro-scopy with a description of pancreas and liver. *J. Cell. Biol.*, **51**, 772–804.

Christensen, A. K., Komorowski, T. E., Wilson, B., Ma, S.-F., and Stevens, R. W. (1985). The distribution of serum albumin in the rat testis, studied by electron microscope immunocytochemistry on ultrathin frozen sections. *Endocrinology*, **116**, 1983–96.

Coetzee, J. and Van der Merwe, C. F. (1984). Extraction of substances during glutaraldehyde fixation of plant cells. *J. Microsc.*, **135**, 147–58.

Cosslett, V. E. (1978). Radiation damage in the high resolution microscopy of bio-logical materials: a review. *J. Microsc.*, **113**, 113–29.

Costello, M. J. and Corless, J. M. (1978). The direct measurement of temperature changes within freeze-fracture specimens during rapid quenching in liquid coolants. *J. Microsc.*, **112**, 17–37.

Costello, M. J., Fetter, R., and Höchli, M. (1982). Simple procedures for evaluating the cryofixation of biological samples. *J. Microsc.*, **125**, 125–36.

Danscher, G. and Norgaard, J. O. R. (1983). Light microscopic visualization of colloidal gold on resin-embedded tissue. *J. Histochem. Cytochem.*, **31**, 1394–8.

Dollhopf, F. L. and Sitte, H. (1969). Die Shandon-Reichert-Kühleinrichtung FC-150 zum Herstellen von Ultradünn- und Feinschnitten bei extrem niederen Temperaturen. *Mikroskopie*, **25**, 17–32.

Dubochet, J. and Lepault, J. (1984). Cryo-electron microscopy of vitrified water. *Journal de Physique*, **45**, C7/85–94.

Dubochet, J. and McDowall, A. W. (1981). Vitrification of pure water for electron microscopy. *J. Microsc.*, **124**, RP3–4.

Dubochet, J. and McDowall, A. W. (1984*a*). Frozen hydrated sections. In *The science of biological specimen preparation* (ed. J.-P. Revel, T. Barnard, G. H. Haggis), pp. 147–52. SEM Inc., AMF O'Hare, Chicago.

Dubochet, J. and McDowall, A. W. (1984*b*). Cryoultramicrotomy: study of ice crystals and freezing damage. *Proc. 8th Europ. Congr. Electron Microsc.*, Vol. 2, pp. 1407–10, Budapest.

Dubochet, J., Groom, M., and Mueller-Neuteboom, S. (1982*a*). The mounting of macromolecules for electron microscopy with particular reference to surface phenomena and the treatment of support films by glow discharge. In *Advances in optical and electron microscopy*, Vol. 8 (ed. R. Barer, V. E. Cosslett), pp. 107–35. Academic Press, London.

Dubochet, J., Lepault, J., Freeman, R., Berriman, J. A., and Homo, J.-C. (1982*b*). Electron microscopy of frozen water and aqueous solutions. *J. Microsc.*, **128**, 219–37.

Dubochet, J., Chang, J.-J., Freeman, R., Lepault, J., and McDowall, A. W. (1982*c*). Frozen aqueous suspensions. *Ultramicroscopy*, **10**, 55–62.

Dubochet, J., Adrian, M., and Vogel, R. H. (1983*a*). Amorphous solid water obtained by vapour condensation or by liquid cooling: a comparison in the electron microscope. *Cryo-Letters*, **4**, 233–40.

Dubochet, J., McDowall, A. W., Menge, B., Schmid, E. N., and Lickfeld, K. G. (1983*b*). Electron microscopy of frozen-hydrated bacteria. *J. Bact.*, **155**, 381–90.

Dubochet, J., Adrian, M., Lepault, J., and McDowall, A. W. (1985). Cryo-electron microscopy of vitrified biological specimens. *TIBS*, April, pp. 143–6.

Dubochet, J., Adrian, M., Chang, J.-J., Homo, J.-C., Lepault, J., McDowall, A. W., and Schultz, P. (1988). Cryo-electron microscopy of vitrified specimens. *Q. Rev. Biophys.*, **21**, 129–228.

Ebersold, H. R., Lüthy, P., Cordier, J. L., and Müller, M. (1981). A freeze-substitution and freeze-fracture study of bacterial spore structures. *J. Ultrastruct. Res.*, **76**, 71–81.

Echlin, P. (1985). *J. Microsc.*, **140**, 1.

Edelmann, L. (1978). A simple freeze-drying technique for preparing biological tissue without chemical fixation for electron microscopy. *J. Microsc.*, **112**, 243–8.

Edelmann, L. (1986). Freeze-dried embedded specimens for biological micro-analysis. *Scanning Electron Microsc.*, **IV**, 1337–56.

Egerton, R. F. (1980). Chemical measurement of radiation damage in organic samples. *Ultramicroscopy*, **5**, 521–3.

Elder, H. Y., Gray, C. C., Jardine, A. G., Chapman, J. N., and Biddlecombe, W. H. (1982). Optimum conditions for the cryo-quenching of small tissue blocks in liquid coolants. *J. Microsc.*, **126**, 45–61.

Eränko, O. (1954). Quenching of tissues for freeze-drying. *Acta. Anat.*, **22**, 331–6.

Escaig, J. (1982). New instruments which facilitate rapid freezing at 83 K and 6 K. *J. Microsc.*, **126**, 221–9.

Eusemann, R., Rose, H., and Dubochet, J. (1982). Electron scattering in ice and organic materials. *J. Microsc.*, **128**, 239–49.

Faulk, W. P. and Taylor, G. M. (1971). An immunocolloid method for the electron microscope. *Immunochemistry*, **8**, 1081–3.

Fernández-Morán, H. (1952). Application of the ultrathin freezing-sectioning technique to the study of cell structures with the electron microscope. *Arch. Fisik*, **4**, 471–91.

Fernández-Morán, H. (1957). Electron microscopy of nervous tissue. In *Metabolism of the nervous system* (ed. D. Richter), pp. 1–34. Pergamon Press, Oxford.

Franks, F. (1978). Biological freezing and cryofixation. In *Low temperature biological microscopy and microanalysis* (ed. P. Echlin, B. Ralph, E. R. Weibel), pp. 3–16. Royal Microscopical Society, Oxford.

Franks, F. (1983). *Water*. Royal Society of Chemistry, London.

Frederik, P. M. and Busing, W. M. (1986). Cryo-transfer revised. *J. Microsc.*, **144**, 215–21.

Frederik, P. M., Busing, W. M., and Persson, A. (1982). Concerning the nature of the cryosectioning process. *J. Microsc.*, **125**, 167–75.

Geymayer, W., Grasenick, F., and Hödl, Y. (1978). Stabilizing ultrathin cryosections by freeze-drying. *J. Microsc.*, **112**, 39–46.

Glaeser, R. M. (1975). Radiation damage and biological electron microscopy. In *Physical aspects of electron microscopy and microbeam analysis* (ed. B. M. Siegel, D. R. Beaman), pp. 205–29. John Wiley and Sons, New York.

Glaeser, R. M. and Taylor, K. A. (1978). Radiation damage relative to transmission electron microscopy of biological specimens at low temperature: a review. *J. Microsc.*, **112**, 127–38.

Glauert, A. M. (1974). *Fixation, dehydration and embedding of biological specimens: practical methods in electron microscopy*, Vol. 3 (ed. A. M. Glauert). North-Holland Publishing Company, Amsterdam, New York, Oxford.

Glauert, A. M. and Young, R. D. (1989). The control of temperature during polymerization of Lowicryl K4M: there is a low-temperature embedding method. *J. Microsc.*, **154**, 101–13.

Green, S. A., Zimmer, K.-P., Griffiths, G., and Mellman, I. (1987). Kinetics of intracellular transport and sorting of lysosomal membrane and plasma membrane proteins. *J. Cell Biol.*, **105**, 1227–40.

Griffiths, G. and Hoppeler, H. (1986). Quantitation in immunocytochemistry: correlation of immunogold labelling to absolute number of membrane antigens. *J. Histochem. Cytochem.*, **34**, 1389–98.

Griffiths, G., Simons, K., Warren, G., and Tokuyasu, K. T. (1983). Immunoelectron microscopy using thin, frozen sections: application to studies of the intracellular transport of semliki forest virus spike glycoproteins. *Meth. Enzymol.*, **96**, 435–50.

Griffiths, G., McDowall, A. W., Back, R., and Dubochet, J. (1984). On the preparation of cryosections for immunocytochemistry. *J. Ultrastruct. Res.*, **89**, 65–78.

Griffiths, G., Hoflack, B., Simons, K., Mellman, I., and Kornfeld, S. (1988). The mannose 6-phosphate receptor and the biogenesis of lysosomes. *Cell*, **52**, 329–41.

Griffiths, G., Fuller, S. D., Back, R., Hollinshead, M., Pfeiffer, S., and Simons, K. (1989). The dynamic nature of the golgi complex. *J. Cell Biol.*, **108**, 277–97.

Griffiths, G., Back, R., and Marsh, M. (1990). A quantitative analysis of the endocytotic pathway in BHK cells. *J. Cell Biol.* (in press).

Gupta, B. L., Berridge, M. J., Hall, T. A., and Moreton, R. B. (1978). Electron microprobe and ion-selective microelectrode studies of fluid secretion in the salivary glands of *Calliphora*. *J. Exp. Biol.*, **72**, 261–4.

Hagler, H. K. and Buja, L. M. (1984). New techniques for the preparation of thin freeze-dried cryosections for X-ray microanalysis. In *The science of biological specimen preparation* (ed. J.-P. Revel, T. Barnard, G. H. Haggis), pp. 161–6. SEM Inc., AMF O'Hare, Illinois.

Hagler, H. K. and Buja, L. M. (1986). Effect of specimen preparation and section transfer techniques on the preservation of ultrastructure, lipids and elements in cryosections. *J. Microsc.*, **141**, 311–17.

Hagler, H. K., Lopez, L. E., Flores, J. S., Lundswick, R. J., and Buja, L. M. (1983). Standards for quantitative energy-dispersive X-ray microanalysis of biological cryosections: validation and application to studies of the myocardium. *J. Microsc.*, **131**, 221–34.

Hall, T. A. (1979). Biological X-ray microanalysis. *J. Microsc.*, **117**, 145–63.

Hall, T. A. (1986). Properties of frozen sections relevant to quantitative micro-analysis. *J. Microsc.*, **141**, 319–28.

Hall, T. A. and Gupta, B. L. (1979). EDS quantification and application to biology. In *Introduction to analytical electron microscopy* (ed. J. J. Hren, J. I. Goldstein, D. C. Joy), pp. 169–97. Plenum Press, New York.

Hall, T. A. and Gupta, B. L. (1982). Quantification for the X-ray microanalysis of cryosections. *J. Microsc.*, **126**, 333–45.

Hall, T. A. and Gupta, B. L. (1983). The localization and assay of chemical elements by microprobe methods. *Q. Rev. Biophys*, **16**, 279–339.

Hall, T. A. and Gupta, B. L. (1984). The application of EDXS to the biological sciences. *J. Microsc.*, **136**, 193–203.

Halloran, B. P., Kirk, R. G., and Spurr, A. R. (1978). Quantitative electron probe microanalysis of biological thin sections: the use of STEM for measurement of local mass thickness. *Ultramicroscopy*, **3**, 175–84.

Hardy, W. B. (1899). *Journal of physiology*, **24**, 288.

Harvey, D. M. R. (1982). Freeze-substitution. *J. Microsc.*, **127**, 209–21.

Hayat, M. A. (1981). *Fixation for electron microscopy*. Academic Press, New York.

Hayward, S. B. and Glaeser, R. M. (1980). High resolution cold stage for the JEOL 100B and 100C electron microscopes. *Ultramicroscopy*, **5**, 3–8.

Hearse, D. J., Yellon, D. M., Chapell, D. A., Wyse, R. K. H., and Ball, G. R. (1981). A high velocity impact device for obtaining multiple, contiguous, myocardial biopsies. *J. Mol. Cell. Cardiol.*, **13**, 197–206.

Heide, H. G. (1982). Design and operation of cold stages. *Ultramicroscopy*, **10**, 125–54.

Heide, H. G. (1984). Observations on ice layers. *Ultramicroscopy*, **14**, 271–8.

Heuser, J. E., Reese, T. S., Dennis, M. J., Jan, Y., Jan, L., and Evans, L. (1979). Synaptic vesicle exocytosis captured by quick freezing and correlated with quantal transmitter release. *J. Cell Biol.*, **81**, 275–300.

Hillier, I. and Gettner, M. E. (1950). Sectioning of tissue for electron microscopy. *Science*, **112**, 520–3.

Hodson, S. and Marshal, J. (1970). Ultramicrotomy—a technique for cutting ultrathin sections of unfixed frozen biological tissues for electron microscopy. *J. Microsc.*, **91**, 105–17.

Holgate, C. S., Jackson, P., Cowen, P. N., and Bird, C. C. (1983). Immunogold-silver staining: new method of immunostaining with enhanced sensitivity. *J. Histochem. Cytochem.*, **31**, 938–44.

Horrisberger, M. (1984). Electron-opaque markers: a review. In *Immunolabelling for electron microscopy* (ed. J. M. Polak, I. M. Varndell), pp. 17–26. Elsevier Science Publishers, Amsterdam.

Horrisberger, M. and Clerc, M.-F. (1985). Labelling of colloidal gold with protein A. A quantitative study. *Histochemistry*, **82**, 219–23.

Horrisberger, M. and Rosset, J. (1977). Colloidal gold, a useful marker for trans-mission and scanning electron microscopy. *J. Histochem. Cytochem.*, **25**, 295–305.

Hosoi, J., Oikawa, T., Inoue, M., Kokubo, Y., and Hama, K. (1981). Measurement of partial specific thickness (net thickness) of critical-point-dried cultured fibroblasts by energy analysis. *Ultramicroscopy*, **7**, 147–54.

Humbel, B. M. and Müller, M. (1986). Freeze substitution and low temperature embedding. In *The science of biological specimen preparation* (ed. M. Müller, R. P. Becker, A. Boyde, J. J. Wolosewick), pp. 175–83. SEM Inc., AMF O'Hare, Illinois.

Humbel, B. M., Marti, T., and Müller, M. (1983). Improved structural preservation by combining freeze-substitution and low temperature embedding. *Beitr. Elektronen-mikrosk. Direktabb. Oberfl.*, **16**, 585–94.

Ingram, M. J. and Ingram, F. D. (1984). Influences of freeze-drying and plastic embedding on electrolyte distributions. In *The science of biological specimen prep-aration* (ed. J.-P. Revel, T. Barnard, G. H. Haggis), pp. 167–74. SEM Inc., AMF O'Hare, Illinois.

Ingram, M. J. and Ingram, F. D. (1986). Cell volume regulation studies with the electron microprobe. In *The science of biological specimen preparation for micro-scopy and microanalysis* (ed. M. Mueller, R. P. Becker, A. Boyde, J. J. Wolosewick), pp. 43–9. SEM Inc., AMF O'Hare, Illinois.

International Experimental Study Group (1986). Cryoprotection in electron micro-scopy. *J. Microsc.*, **141**, 385–91.

Isaacson, M. S. (1975). Inelastic scattering and beam damage of biological molecules. In *Physical aspects of electron microscopy and microbeam analysis* (ed. B. M. Siegel, D. R. Beaman), pp. 247–58. John Wiley and Sons, New York.

Isaacson, M. S. (1977). Specimen damage in the electron microscope. In *Principles and techniques of electron microscopy, biological applications*, Vol. 7 (ed. M. A. Hayat), pp. 1–78. Van Nostrand Reinhold Co., New York.

Johansen, B. V. (1973). Bright field electron microscopy of biological specimens: 1. Obtaining the optimum contribution of phase contrast to image formation. *Micron*, **4**, 446–72.

Karp, R. D., Silcox, J. C., and Somlyo, A. V. (1982). Cryoultramicrotomy: evidence against melting and the use of a low temperature cement for specimen orientation. *J. Microsc.*, **125**, 157–65.

Kendall, M. D., Warley, A., and Morris, I. W. (1985). Differences in apparent elemental composition of tissues and cells using a fully quantitative X-ray micro-analysis system. *J. Microsc.*, **138**, 35–42.

Larsson, L.-I. (1979). Simultaneous ultrastructural demonstration of multiple peptides in endocrine cells by a novel immunocytochemical method. *Nature*, **282**, 743–6.

Latta, H. and Hartmann, J. F. (1950). Use of glass edge in thin sectioning for electron microscopy. *Proc. Soc. Exp. Biol. Med.*, **74**, 436–9.

Leapman, R. D., Fiori, C. E., and Swyt, C. R. (1984). Mass thickness determination by electron energy loss for quantitative X-ray microanalysis in biology. *J. Microsc.*, **133**, 239–53.

Leathem, A. (1986). Lectin histochemistry. In *Immunocytochemistry, modern methods and applications* (2nd edn) (ed. J. M. Polak, S. van Noorden), pp. 167–87. John Wright and Sons, Bristol.

Lepault, J., Booy, F. P., and Dubochet, J. (1983). Electron microscopy of frozen biological suspensions. *J. Microsc.*, **129**, 89–102.

Linders, P. W. J. and Hagemann, P. (1983). Mass determination of thin biological specimens using backscattered electrons. Application in quantitative X-ray microanalysis on an automated STEM system. *Ultramicroscopy*, **11**, 13–20.

Linders, P. W. J., Stols, A. L. H., van der Vorstenbosch, R. A., and Stadhouders, A. M. (1982). Mass determination of thin biological specimens for use in quantitative electron probe X-ray microanalysis. *Scan. Electron Microsc.*, **4**, 1603–15.

Linders, P. W. J., van der Vorstenbosch, R. A., Smits, H. T. J., Stols, A. L. H., and Stadhouders, A. M. (1984). Absolute quantitative electron microscopy of thin biological specimen by energy-dispersive X-ray microanalysis and densitometric mass determination. *Anal. Chim. Acta*, **160**, 57–67.

Marshall, A. T. (1980). Freeze-substitution as a preparation technique for biological X-ray microanalysis. *Scanning Electron Microsc.* **II**, 395–408.

Marshall, A. T. (1988). Progress in quantitative X-ray microanalysis of frozen-hydrated bulk biological samples. *J. Electron Microsc. Tech.*, **9**, 57–64.

McDowall, A. W., Chang, J.-J., Freeman, R., Lepault, J., and Walter, C. A. (1983). Electron microscopy of frozen hydrated sections of vitreous ice and vitrified biological samples. *J. Microsc.*, **131**, 1–9.

Meek, G. A. (1976). *Practical electron microscopy for biologists* (2nd edn). John Wiley and Sons, London.

Milligan, R. A., Brisson, A., and Unwin, P. N. T. (1984). Molecular structure determination of crystalline specimens in frozen aqueous solutions. *Ultramicroscopy*, **13**, 1–9.

Monroe, R. G., Gamble, W. J., La Farge, C. G., Gamboa, R., Morgan, C. L., Rosenthal, A., and Bullivant, S. (1968). Myocardial ultrastructure in systole and diastole using ballistic cryofixation. *J. Ultrastruct. Res.*, **22**, 22–36.

Moor, H. (1987). Theory and practice of high pressure freezing. In *Cryotechniques in biological electron microscopy* (ed. R. A. Steinbrecht, K. Zierold), pp. 175–91. Springer Verlag, Berlin, Heidelberg.

Moor, H., Kristler, I., and Müller, M. (1976). Freezing in a propane jet. *Experientia*, **32**, 805.

Morgan, A. J. (1980). Preparation of specimens. Changes in chemical integrity. In *X-ray microanalysis in biology* (ed. M. A. Hayat), pp. 65–165. University Park Press, Baltimore.

Morgan, A. J. (1985). *X-ray microanalysis in electron microscopy for biologists.* Royal Microscopical Society/Oxford University Press, Oxford.

Morris, G. J. (1981). *Cryopreservation: an introduction to cryopreservation in culture collections.* Institute of Terrestrial Ecology, Cambridge.

Morris, G. J. and Clarke, A. (ed.) (1981). *Effects of low temperatures on biological membranes.* Academic Press, London.

Müller, M. (1981). Demonstration of liposomes by electron microscopy. In *Membrane proteins* (ed. A. Azzi, U. Brodbeck, Z. Zahler), pp. 252–6. Springer Verlag, Heidelberg.

Müller, M., Meister, N., and Moor, H. (1980). Freezing in a propane jet and its application in freeze-fracturing. *Mikroskopie*, **36**, 129–40.

Namork, E. and Johansen, B. V. (1982). Surface activation of carbon supports for biological electron microscopy. *Ultramicroscopy*, **7**, 321–30.

Needham, J. (1936). *Order and life.* Cambridge University Press, Cambridge.

Newbury, D. E., Joy, D. C., Echlin, P., Fiori, C. E., and Goldstein, J. I. (1986). *Advanced scanning electron microscopy and X-ray microanalysis.* Plenum Press, New York.

Newman, G. R. and Hobot, J. A. (1987). Modern acrylics for post-embedding immunostaining techniques. *J. Histochem. Cytochem.*, **35**, 971–81.

Newman, S. B., Borysko, E., and Swerdlow, M. (1949). Ultramicrotomy by a new method. *I. Res. Nat. Bureau of Standards*, **43**, 183–99.

Newman, G. R., Jasani, B., and Williams, E. D. (1982). The preservation of ultrastructure and antigenicity. *J. Microsc.*, **127**, RP5–6.

Newman, G. R., Jasani, B., and Williams, E. D. (1983). A simple postembedding system for the rapid demonstration of tissue antigens under the electron microscope. *Histochem. J.*, **15**, 543–55.

Nicholson, W. A. P., Biddlecome, W. H. and Elder, H. Y. (1982). Low X-ray background low temperature specimen stage for biological microanalysis in the transmission electron microscope. *J. Microsc.*, **126**, 307–16.

Nott, J. A. and Langston, W. J. (1989). Cadmium and the phosphate granules in *Littorina littorea. J. Mar. Biol. Ass.*, **69**, 219–27.

O'Brien, H. C. and McKinley, G. M. (1943). New microtome and sectioning method for electron microscopy. *Science*, **98**, 455–6.

Parsons, D., Bellotto, D. J., Schulz, W. W., Buja, M., and Hagler, H. K. (1984). Towards routine cryoultramicrotomy. *EMSA Bulletin*, **14**, 49–60.

Perlov, T., Talmon, Y., and Falls, A. H. (1983). An improved transfer module and variable temperature control for a simple commercial cooling holder. *Ultramicroscopy*, **11**, 283–8.

Philips, T. E. and Boyne, A. F. (1984). Liquid nitrogen-based quick freezing: experiences with bounce-free delivery of cholinergic nerve terminals to a metal surface. *J. Electron Microsc. Tech.*, **1**, 9–29.

Plattner, H. and Bachmann, L. (1982). Cryofixation: a tool in biological ultrastructural research. *Int. Rev. Cytol.*, **79**, 237–304.

Plattner, H. and Knoll, G. (1987). Ultrastructural analysis of dynamic cellular processes: a survey of current problems, pitfalls and perspectives. *Scanning microscopy*, **1**, 1199–216.

Plattner, H., Smitt-Fumian, W. W., and Bachmann, L. (1973). Cryofixation of single cells by spray-freezing. In *Freeze-etching. Techniques and applications* (ed. E. L. Benedetti, P. Farards), pp. 81–100. Soc. Francaise de Microscopie Electronique, Paris.

Posthuma, G., Slot, J. W., and Geuze, H. J. (1987). Usefulness of the immunogold technique in quantitation of a soluble protein in ultrathin sections. *J. Histochem. Cytochem.*, **35**, 405–10.

Pscheid, P., Schudt, C., and Plattner, H. (1981). Cryofixation of monolayer cell cultures for freeze-fracturing without chemical pretreatments. *J. Microsc.*, **121**, 149–67.

Rasmussen, D. H. (1982). Ice formation in aqueous systems. *J. Microsc.*, **128**, 167–74.

Reimer, L. (1975). Review of the radiation damage problem of organic specimens in electron microscopy. In *Physical aspects of electron microscopy and microbeam analysis* (ed. B. M. Siegel, D. R. Beaman), pp. 231–45. John Wiley and Sons, New York.

Rick, R., Dörge, A., Gehring, K., Bauer, R., and Thurau, K. (1979). Quantitative determination of cellular electrolyte concentrations in thin freeze-dried cryosections using energy dispersive X-ray microanalysis. In *Microbeam analysis in biology* (ed. C. P. Lechene, R. R. Warner), pp. 517–34. Academic Press, New York.

Rick, R., Dörge, A., and Thurau, K. (1982). Quantitative analysis of electrolytes in frozen-dried sections. *J. Microsc.*, **125**, 239–47.

Robards, A. W. and Sleytr, U. B. (1985). *Low temperature methods in biological electron microscopy: practical methods in electron microscopy*, Vol. 10 (ed. A. M. Glauert). Elsevier, Amsterdam, New York, Oxford.

Roberts, I. M. (1975). Tungsten coating—a method of improving glass microtome knives for cutting ultrathin sections. *J. Microsc.*, **103**, 113–19.

Romano, E. L. and Romano, M. (1977). Staphylococcal protein A bound to colloidal gold. A useful reagent to label antigen–antibody sites in electron microscopy. *Immunochemistry*, **14**, 711–15.

Romano, E. L., Stolinski, C., and Hughes-Jones, N. C. (1974). An antiglobulin reagent labelled with colloidal gold for use in electron microscopy. *Immunochemistry*, **11**, 521–2.

Roos, N. and Barnard, T. (1984). Aminoplastic standards for quantitative X-ray microanalysis of thin sections of plastic embedded biological material. *Ultramicroscopy*, **15**, 277–86.

Roos, N. and Barnard, T. (1985). A comparison of subcellular element concentrations in frozen-dried, plastic-embedded, dry-cut sections and frozen-dried cryosections. *Ultramicroscopy*, **17**, 335–44.

Roos, N. and Barnard, T. (1986). Preparation methods for quantitative electron probe X-ray microanalysis of rat exocrine pancreas: a review. *Scanning Electron Microsc.*, **2**, 703–11.

Roos, N. and Morgan, A. J. (1985). Aminoplastic standard for electron probe X-ray microanalysis (EPXMA) of ultrathin frozen-dried cryosections. *J. Microsc.*, **140**, RP3–4.

Roos, N., Kinde, U., and Morgan, J. (1990). The morphology of rat exocrine pancreas prepared by anhydrous cryo-procedures. *J. Electron Microsc. Tech.*, **14**, 39–45.

Roth, J. (1983). The colloidal gold marker system for light and electron microscopy cytochemistry. In *Techniques in immunocytochemistry II* (ed. G. R. Bullock, P. Petrusz), pp. 217–84. Academic Press, London and New York.

Roth, J., Bendayan, M., and Orci, L. (1978). Ultrastructural localization of intracellular antigens by the use of protein A–gold complex. *J. Histochem. Cytochem.*, **26**, 1074–81.

Roth, J., Bendayan, M., Carlemalm, E., Villiger, W., and Garavito, M. (1981). Enhancement of structural preservation and immunocytochemical staining in low temperature embedded pancreatic tissue. *J. Histochem. Cytochem.*, **29**, 663–71.

Ryan, K. P. and Liddicoat, M. I. (1987). Safety considerations regarding the use of propane and other liquefied gases as coolants for rapid freezing purposes. *J. Microsc.*, **147**, 337–40.

Ryan, K. P., Purse, D. H., Robinson, S. G., and Wood, J. W. (1987). The relative efficiency of cryogens used for plunge-cooling biological specimens. *J. Microsc.,* **145**, 89–96.

Saubermann, A. J. and Heymann, R. V. (1987). Quantitative digital X-ray imaging using frozen-hydrated and frozen-dried tissue sections. *J. Microsc.,* **146**, 169–82.

Saubermann, A. J., Riley, W. D., and Beeuwkes, R. (1977). Cutting work in thick section cryomicrotomy. *J. Microsc.,* **111**, 39–49.

Shuman, H., Somlyo, A. V., and Somlyo, A. P. (1976). Quantitative electron probe microanalysis of biological thin sections: methods and validity. *Ultramicroscopy,* **1**, 317–39.

Shuman, H., Somlyo, A. V., and Somlyo, A. P. (1977). Theoretical and practical limits of ED X-ray analysis of biological thin sections. *Scanning Electron Microsc.,* **1**, 663–72.

Sitte, H. (1984). Equipment for cryofixation, cryoultramicrotomy and cryosubstitution in bio-medical TEM-routines. *Zeiss information magazine for electron microscopists,* **3**, 25–31.

Skaer, H. (1982). Chemical cryoprotection for structural studies. *J. Microsc.,* **125**, 137–47.

Slot, J. W. and Geuze, H. J. (1981). Sizing of protein A–colloidal gold probes for immunoelectron microscopy. *J. Cell Biol.,* **90**, 533–6.

Slot, J. W. and Geuze, H. J. (1985). A new method of preparing gold probes for multiple-labelling cytochemistry. *Eur. J. Cell Biol.,* **38**, 87–93.

Slot, J. W., Geuze, J. H., Freeman, B. A., and Crapo, J. D. (1986). Intracellular localization of the copper–zinc and manganese superoxide dismutases in rat liver parenchymal cells. *Lab. Invest.,* **55**, 363–71.

Somlyo, A. V., Shuman, H., and Somlyo, A. P. (1977). Elemental distribution in striated muscle and the effects of hypertonicity: electron probe analysis of cryo sections. *J. Cell Biol.,* **74**, 828–57.

Somlyo, A. P., Bond, M., and Somlyo, A. V. (1985). Calcium content of mitochondria and endoplasmic reticulum in liver frozen rapidly *in vivo. Nature,* **314**, 622–5.

Stang, E. (1988). Modification of the LKB 7800 series knife maker for symmetrical breaking of 'cryo' knives. *J. Microsc.,* **149**, 77–9.

Steinbrecht, R. A. (1985). Recrystallization and ice-crystal growth in a biological specimen, as shown by a simple freeze-substitution method. *J. Microsc.,* **140**, 41–6.

Steinbrecht, A. and Zierold, K. (ed.) (1987). *Cryotechniques in biological electron microscopy.* Springer Verlag, Heidelberg.

Stephenson, J. L. (1956). Ice crystal growth during the rapid freezing of tissues. *J. Biophysic. Biochem. Cytol.,* **2**, 45–52.

Stephenson, J. L. (1960). Ice crystal formation in biological materials during rapid freezing. *Ann. NY Acad. Sci.,* **85**, 535–40.

Stirling, C. E. and Kinter, W. B. (1967). High resolution radioautography of galactose-^{3}H accumulation in rings of hamster intestine. *J. Cell Biol.,* **35**, 585–604.

Stillinger, F. H. (1980). Water revisited. *Science,* **209**, 451–7.

Storey, K. B. and Storey, J. M. (1988). Freeze tolerance in animals. *Physiol. Rev.,* **68**, 27–84.

Stumpf, W. E. and Roth, L. J. (1967). Freeze-drying of small tissue samples and thin frozen sections below −60 °C. *J. Histochem. Cytochem.,* **15**, 243–51.

Talmon, Y., Adrian, M., and Dubochet, J. (1986). Electron beam radiation damage to organic inclusions in vitreous, cubic and hexagonal ice. *J. Microsc.*, **141**, 375–84.

Taylor, K. A. (1978). Structure determinations of frozen, hydrated, crystalline biological specimens. *J. Microsc.*, **112**, 115–25.

Taylor, K. A. and Glaeser, R. M. (1974). Electron diffraction of frozen, hydrated protein crystals. *Science*, **186**, 1036–7.

Tokuyasu, K. T. (1973). A technique for ultramicrotomy of cell suspensions and tissues. *J. Cell Biol.*, **57**, 551–65.

Tokuyasu, K. T. (1986*a*). Application of cryoultramicrotomy to immunocytochemistry. *J. Microsc.*, **143**, 139–49.

Tokuyasu, K. T. (1986*b*). Cryosections for immunohistochemistry. *Proc. XIth Internat. Congr. EM*, Kyoto, Japan, 42–3.

Unwin, P. N. T. and Henderson, R. (1975). Molecular structure determination by electron microscopy of unstained crystalline specimens. *J. Molec. Biol.*, **94**, 425–40.

Umrath, W. (1983). Berechnung von Gefriertrocknungszeiten für die elektronenmikroskopische Präparation. *Mikroskopie*, **40**, 9–37.

Valdrè, U. (1979). Electron microscope stage design and applications. *J. Microsc.*, **117**, 55–75.

Van Harreveld, A., Trubatch, J., and Steiner, J. (1974). Rapid freezing and electron microscopy for the arrest of physiological processes. *J. Microsc.*, **100**, 189–98.

Varndell, I. M. and Polak, J. M. (1987). EM immunolabelling. In *Electron microscopy in molecular biology* (ed. J. Sommerville, U. Scheer), pp. 179–200. IRL Press, Oxford.

Von Zglinicki, T. and Bimmler, M. (1987). The intracellular distribution of ions and water in rat liver and heart muscle. *J. Microsc.*, **146**, 77–85.

Von Zglinicki, T., Bimmler, M., and Krause, W. (1987). Estimation of organelle water fractions from frozen-dried cryosections. *J. Microsc.*, **146**, 67–75.

Von Zglinicki, T., Bimmler, M., and Purz, H.-J. (1986). Fast cryofixation technique for X-ray microanalysis. *J. Microsc.*, **141**, 79–90.

Warner, R. R. (1986). Water content from analysis of freeze-dried thin sections. *J. Microsc.*, **142**, 363–9.

Warley, A. (1986). Elemental concentrations in isolated rat thymocytes prepared for cryofixation in the presence of different media. *J. Microsc.*, **144**, 183–91.

Webster, P., Dobbelaere, D. A. E., and Fawcett, D. W. (1985). The entry of sporozoites of *Theileria parva* into bovine lymphocytes *in vitro*. Immunoelectron microscopic observations. *Eur. J. Cell Biol.*, **36**, 157–62.

Wendt-Gallitelli, M.-F., and Wolburg, H. (1984). Rapid freezing, cryosectioning, X-ray microanalysis on muscle preparations in defined functional states. *J. Electron Microsc. Tech.*, **1**, 151–74.

Williams, R. C. and Fisher, H. W. (1970). Electron microscopy of tobacco mosaic virus under conditions of minimal beam exposure. *J. Molec. Biol.*, **52**, 121–3.

Wolfgang, W. J., Fristrom, D., and Fristrom, J. W. (1986). The pupal cuticle of *Drosophila*: differential ultrastructural immunolocalization of cuticle proteins. *J. Cell Biol.*, **102**, 306–11.

Wrigley, N. G., Brown, E., and Chillingworth, R. K. (1983). Combining accurate defocus with low-dose imaging in high resolution electron microscopy of biological material. *J. Microsc.*, **130**, 225–32.

Wroblewski, J. and Wroblewski, R. (1986). Why low temperature embedding for X-ray microanalytical investigations? A comparison of recently used preparation methods. *J. Microsc.*, **142**, 351–62.

Wroblewski, R., Wroblewski, J., Anniko, M., and Edström, L. (1985). Freeze-drying and related preparation techniques for biological microprobe analysis. *Scanning Electron Microsc.,* **I**, 447–54.

Wroblewski, J., Wroblewski, R., and Roomans, G. M. (1988). Low temperature techniques for X-ray microanalysis in pathology: alternatives to cryoultramicrotomy. *J. Electron Microsc. Tech.,* **9**, 83–98.

Wynford-Thomas, D. Jasani, B., and Newman, G. R. (1986). Immunocytochemical localization of cell surface receptors using a novel method permitting simple, rapid, and reliable LM/EM correlation. *Histochem. J.,* **18**, 387–96.

Zasadzinski, J. A. N. (1988). A new heat transfer model to predict cooling rates for rapid freezing fixation. *J. Microsc.,* **150**, 137–49.

Zierold, K. (1982). Preparation of biological cryosections for analytical electron microscopy. *Ultramicroscopy,* **10**, 45–54.

Zierold, K. (1984). The morphology of ultrathin cryosections. *Ultramicroscopy,* **14**, 201–10.

Zierold, K. (1986). The determination of wet weight concentrations of elements in freeze-dried cryosections from biological cells. *Scanning Electron Microsc.,* **2**, 713–24.

Zierold, K. (1988). X-ray microanalysis of freeze-dried and frozen-hydrated cryosections. *J. Electron Microsc. Tech.,* **9**, 65–82.

Zingsheim, H. P. and Plattner, H. (1976). Electron microscopy methods in biology. In *Methods in membrane biology* Vol. 7 (ed. E. D. Korn), pp. 1–146. Plenum Publishing Company, New York.

Index